GENERATION ROBOT

GENERATION ROBOT

A Century of Science Fiction, Fact, and Speculation

Terri Favro

Skyhorse Publishing

10 9 8 7 6 5 4 3 2 1

Library of Congress Cataloging-in-Publication Data is available on file.

Cover design by Erin Seaward-Hiatt
Cover Image: IStock

Print ISBN: 978-1-5107-2310-8
Ebook ISBN: 978-1-5107-2312-2

Printed in the United States of America

In memory of Attilio "Tee" Favro

Roboticist, inventor, father

CONTENTS

“I sometimes think that, in the desperate straits of humanity today, we would be grateful to have nonhuman friends, even if they are only friends we build ourselves.”

—Isaac Asimov

Introduction

WHY ROBOTS?

Once, they promised us the moon . . .

We grew up knowing that nuclear bombs could destroy Earth, but also that we could escape to other worlds. Seesawing between paranoia and a switched-on, backcombed, Tupperware-partying, wipes-clean-with-a-damp-cloth faith in progress, we anticipated a future that would never arrive—at least, not the one sold to us as the World of Tomorrow.

Lurking in the shadows of these two opposing dreams of the future—one dark and full of visions of doom, the other as bright and cheery as a freshly washed linoleum floor—was a creature who had been part of human imagination for centuries, but who truly came into its own at the dawn of the atomic age: the robot.

We expected robots to be our companions and protectors. Tireless, trustworthy, and indestructible, they would be smarter and stronger than we were, but lacking the human aggression that had taken Earth to the brink of annihilation. One day, they would keep house for us, like Rosey the wisecracking robot maid on The Jetsons, *or explore other planets with us, like Will Robinson's robot on* Lost In Space.

As time passed, and the bomb did not fall—at least, not on us—and the lunar colonies failed to materialize, we became jaded and the World of Tomorrow turned into a sad dream, like a rusty amusement park ride at an abandoned World's Fair.

But even though we never got to live on the moon, we are getting our robots. Now we stand to be among the first generation to hand them our car keys and let them be the Earth's grown-ups. In the words of HAL 9000, the homicidal artificial intelligence from 2001: A Space Odyssey, *it's time for us to "sit down calmly, take a stress pill and think things over."*

To borrow a phrase from *Star Trek*'s Leonard "Bones" McCoy: I'm a writer, not a roboticist. I've spent the last thirty years as a marketing copywriter, lifestyle journalist, humorist, novelist, and blogger. You might ask, then, why I chose to write a book about my generation's relationship with robots, artificial intelligence, and computers.

I have my reasons. Seven of them, actually.

Reason one: I loved my dad and he loved robots. Unlike everyone else in my family, I was weak at math, which deeply troubled my father, an electrician and amateur inventor, and my mom, who worked as a book-keeper before settling down to be a wife and mother. Mom prided herself on being able to add up columns of figures in her head: she would have been a good computer, like the code-breaking women at Bletchley Park who helped defeat Hitler. My sisters Rosemary and Johanna both excelled at math and physics; Rosemary, the oldest, married Roger, an ex-seminarian who left the priesthood to study pure mathematics and become an IBM software developer in 1967. My brother Rick wanted to be an astronaut, but instead studied industrial engineering and went to Silicon Valley in the 1990s to develop Internet-based telephone systems.

Me? I ended up as an advertising copywriter for tech clients including IBM, Apple, early cell phone companies, and Internet banking. I have the dubious distinction of writing marketing copy for one of the biggest tech marketing flops in history, the IBM PCJr.

My comfort with these clients was partly because of their white-shirted, pocket-protected, geeky familiarity: growing up in my lively Italian family, the love of all things tech meant that dinners often ended with everyone sitting around the table, drinking bottle after bottle of my father's homemade wine, and talking about moon shots, cars, computers and, especially robots. That's because for the last decade of his career, Dad worked with the first factory robot, UNIMATE, which inspired him to build robots at home. So, although I'm not a scientist or engineer myself, I grew up around the peculiar creative spirit that fires them up.

Reason two: Robots live deep inside my pop culture DNA. I could have been little Gloria from "Robbie," the first story in Isaac Asimov's story collection, *I, Robot,* playing hide-and-seek with her robot babysitter until her mother banishes him to the drudgery of factory work. In the end, Robbie saves Gloria's life, redeeming himself like a mechanical Pinocchio whose goodness turns him into something approaching a real boy. The moral of the story seemed to be that robots are more humane than humans—our smarter, saner, altruistic step-siblings, the perfect kids every suburban North American household wanted to raise. Robots didn't turn into juvenile delinquents, smoke dope, or end up in homes for wayward girls. And unlike many real kids, Asimov's children never let their parents down.

In the 1960s, I bathed in the glow of TV shows like *Lost in Space* and *My Living Doll,* whose stars were robots. Like every other bookish preteen my age, I never missed an episode of *Star Trek* and developed a hopeless Spock-crush. And what was Spock, if not a template for those next-generation Trekkie androids and cyborgs: Data and Seven Of Nine? When NBC announced the cancellation of *Star Trek,* I joined the protest movement of outraged letter writers. We managed to give the series an extra year of life on the small screen before Captain James T. Kirk was forced into a second career as a margarine pitchman.

In 1968, my sisters took me to see *2001: A Space Odyssey* where I met the most influential robot since Asimov's creations: HAL 9000, an astronaut-murdering, Daisy Bell-singing artificial intelligence rocketing to Jupiter with its human partners. At eleven years of age, I didn't understand why HAL refused to open the pod bay door for Astronaut Dave, or why Dave voyaged through a psychedelic universe to transform into a fetus-like star child. The symbolism was beyond me. But I did start saving up for my own winter-white junior miss futuristic stewardess jumpsuit from the Sears catalogue.

In the seventies, ever-cooler robots entered my life through movies, TV, and books. *Silent Running, Star Wars, The Hitchhiker's Guide to the Galaxy, Alien,* and *Blade Runner* all built on Asimov's vision by introducing newer, often more sinister, sometimes cuter robots, while novelists Philip K. Dick, Stephen King, and Arthur C. Clarke created alternate realities where humans and robots coexisted in an uneasy truce. By the time the IBM PC and Apple Macintosh swept the 1980s like a tsunami, washing away entire job descriptions (remember typesetters?), young adults my age were prepped and ready: we had seen our futures hanging before us like *Blade Runner*'s Harrison Ford dangling from his dislocated fingers while replicant Rutger Hauer clutched a dove and poetically reminisced about attack ships off the shoulder of Orion before slumping over into a preprogrammed death. Technology was the future: adapt or die. So, my generation adapted. Even though we weren't going to get flying cars and jet packs, desktop computers and the Internet were going to make life just as Magical Kingdom-ish right here on good old Earth, if we didn't blow it up or poison it to death first.

Growing up in a state of constant anticipation for the next big thing, we are now seeing the pop cult robots of our childhoods become reality—from the Internet of Things, to automated kitchens, to sex dolls that make AI-driven pillow talk, to humanoid care-giving robots for the children and frail seniors, and a system from Amazon called anticipatory shipping that delivers purchases to us *before* we even know we want them.[1] How cool—or terrifying—is that?

Reason three: Robots are my generation's birthright. I'm a member of so-called Generation Jones, late boomers who were unlucky enough to be stuck between the prosperous, idealistic first wave of boomers and the disillusioned, but cooler, Generation X. Born between 1954 and

1962, we came of age too late for love-ins but just in time for the energy crisis and economic stagflation of the late 1970s. We were the generation always jonesing for the success achieved by our older siblings but never quite able to live up to our potential.

Today we are pigeonholed as *digital immigrants* to distinguish us from our kids and grandkids, digital natives born with a touch screen in their hands. The word *immigrant* implies that we are newcomers to technology, struggling to adapt to an alien world where we speak the language imperfectly.

And yet, we have always been defined by our technology. We were the first cohort with TV babysitters, synthesized music, artificial snacks, and the residue of atmospheric nuclear tests embedded into our baby teeth. A better label for us might be *digital amnesiacs*. We have lived our lives on the rolling deck of a fast-moving technological ship, leaving in our wake a trail of mechanical discards—TVs, fax machines, word processors, videocams, eight-track tapes, VCRs, CD players, desktop computers, laptops, and the earliest brick-shaped cell phones, bobbing, then sinking as we junk our old devices, moving on to the next innovation, and forgetting we lived our lives any other way. Part of my motivation in writing this book is to take stock of the fantastic voyage we've been on throughout our lifetimes.

Reason four: Robots were artistic creations long before they became digital electromechanical ones. Rather than being a natural outgrowth of technological research—like, say, the electric lightbulb or the telegraph—thinking machines came to life as the offspring of novelists, playwrights, and filmmakers, which inspired scientists, engineers, and industrial designers to make real the images conjured up in films and books. Even today, half of the nominees in the Carnegie-Mellon Robot Hall of Fame are imaginary, including R2D2 from *Star Wars*, WALL-E, the garbage compacting robot from the animated Disney film of the same name, and, of course, HAL 9000 of *2001: A Space Odyssey*.[2]

Binge watching *2001* inspires computational linguistics.

High school student Jerry Kaplan saw *2001: A Space Odyssey* six times over the summer of 1968. The film inspired him to announce to his friends that his mission in life was to build his own HAL. He earned his PhD in computer science and became a pioneer in artificial intelligence, specializing in communicating with computers using natural English[3]—the way that astronauts Bowman and Poole conversed with HAL, and the way rest of us ask Siri where to find the nearest Malaysian-Greek fusion hipster restaurant.

Reason five: The true story of robots and AI in our everyday lives still needs to be told (without getting muddled up in clickbait and marketing hype). Computer scientists and engineers have been working to design autonomous robots that could move, learn, and make decisions for far longer than most of us nonscientists realize: the first walking, seeing robot was developed at Stanford in 1970, a mere year after *Star Trek* finally went off the air despite my protests, a full seven years before R2D2 and C3Po appeared in *Star Wars*. Even the self-driving car isn't a new idea: an autonomous vehicle was in development by the mid-1960s as part of the Mercury-Gemini-Apollo space program, until it was mothballed in favor of a lunar rover that would be driven by a human astronaut.[4]

Meanwhile, out of view of mere mortals like you and me, a behind-the-scenes war has been raging for decades in the world of robotics and computing: a showdown between AI (where the ultimate goal is to develop autonomous robots that could replace us outright) and IA (Intelligence Augmentation), in which robots and computers would act as our helpers—or as the Czech word *robot* originally implied, our slaves—rather than our replacements. While many Silicon Valley types expected the AI revolution in the early 1980s, the real revolution turned out to be in personal computing, led in part by IA champion Steve Jobs who viewed the Macintosh personal computer as a bicycle for the human mind, rather than a replacement for it.[5]

But times, again, are a-changing. Powerful AIs like Google and Siri are now part of everyday life. Some of us balance our stock portfolios with help from robot investment advisors and diagnose our illnesses using the algorithms on "Doctor Google"; I've even watched my physician log into MedLine.com and key in my symptoms, something I could do myself without a medical degree. And, of course, the first generation of the much-hyped, look-ma-no-hands, fully autonomous cars from Tesla, Ford, GM, Toyota, BMW, Google, and (it's rumored) Apple, may start appearing in showrooms by 2025.[6] With the exception of the Google car, which looks like a blob of Play-Doh stuck on wheels, or something Clarabelle Cow might drive in a 1930s Looney Tune cartoon, all these driverless cars are sleekly designed road machines with steering wheels visible through the windshields, as if trying to reassure us that we are still in control.

It's not just scientists and engineers who get to decide if robots take over. You do, too. Consumer acceptance matters a lot, because self-driving cars, devices connected to the Internet of Things, and household robots, will be marketed and sold as *products*. We will have to be persuaded to buy in, or these robots will fail simply because we don't like them. So, as consumers of robotics, it's important for us to get past the overblown hype about *Terminator*-like killer robots and gain a clearer view of the impact of these machines, good and bad, on our future lives—which ones are good value for the money, and which are destined to rust away in warehouses.

Reason six: We don't know when robots will become a regular part of everyday life, or what will change for us when they do. After almost sixty years of development, inspiration, sweat, tears, and millions if not billions of dollars in venture capital, robots do not quite yet walk among us (although they do kill our enemies, sweep for land mines, check for bombs, visit other planets, pick our crops, and do a lot of our factory work) but they seem poised to become our benevolent caretakers and perhaps even our accidental gods.

They also have the potential to do considerable harm to humans. The economic fallout from the roboticization of industry, agriculture, banking, the food service industry, shipping, accounting—well, you name the job and there's probably a robot in development to do it—could be staggering. In fact, the impact of robots in the workplace has been affecting job growth since the 1970s[7] but like a lobster that doesn't realize it's boiling to death until it's too late to crawl out of the pot, we're only just starting to wake up to the fact that robots are taking away not only dirty, dangerous jobs—like cleaning up nuclear power plants—but could soon also be performing high-paying, rewarding, and creative ones, like medicine, law, and even music and journalism.

The most disappointing robots are the ones we're secretly longing for the most: the humanoid synthetics that walk and talk and wave their dryer-hose arms in the air warning us of *danger*! The ones that bleed white blood when they're torn in half bravely trying to save a human's life, like Bishop in *Aliens*. Ones that look like us, talk like us, but are smarter, stronger, and nicer than we are, the ultimate big brother who always has our backs. YouTube videos of family-friendly robots that sing lullabies to children, do aerobics with Mom, and greet heads of state at corporate headquarters don't quite live up to the imaginary robots on film and in books. Even at their best and most terrifying, like the Boston Dynamics' robotic dogs and horses, none of them act autonomously—yet. They are all just a little too animatronic. It will take the next step in the evolution of Artificial Intelligence, known as Artificial General Intelligence or AGI, to create a robot that might pass as one of us. What science fiction has promised, the engineers must make good on. Will they succeed?

For argument's sake, let's assume that venture capital keeps flowing, and the roboticists finally catch up with our pop cult imaginations. How soon will we see truly conscious machines of the type in *Blade Runner*, *Her*, and *Ex Machina*? That's a subject of intense debate even among roboticists themselves. In *Rise of the Robots: Technology and the Threat of a Jobless Future*, author and Silicon Valley–based software developer Martin Ford describes a survey by AI writer James Barrat who asked two hundred AGI researchers to predict when we would see a thinking machine:

The results: 42 percent believed a thinking machine would arrive by 2030, 24 percent said by 2050, and 20 percent thought it would happen by 2100. Only 2 percent believed it would never happen. Remarkably, a number of respondents wrote comments on their surveys suggesting that Barrat should have included an even earlier option—perhaps 2020.[8]

Ripley's good and evil artificial people.

Pity poor Ellen Ripley (Sigourney Weaver). Science Officer Ash (Ian Holm) uses her and her shipmates as live bait in *Alien* (1979): turns out he's a robot following orders to do whatever's necessary to bring a malevolent new species back to Earth. "I admire its purity," Ash murmurs while dissecting the disgusting face hugger that deposits the parasitical alien fetus into Executive Officer Kane's (John Hurt) chest. But before we can comfortably assume that every robot in the *Alien* movie series is the sadistic lackey of a soulless corporation, Ripley meets Bishop (Lance Henrickson) in *Aliens* (1986): an altruistic synthetic person (or to use the politically correct term Bishop prefers, "artificial human") who quotes Asimov's First Law when he hears of Ash's sabotage: "I'm shocked! Was it an older model? The A2s always were a bit twitchy. That could never happen now with our behavioral inhibitors. It is impossible for me to harm or by omission of action, allow to be harmed, a human being."

Whether the breakthrough to machine consciousness will come in a hundred years or six months from now, we do know that the pace of change is accelerating. One reason is Moore's Law, named after one of the founders of Intel, Gordon Moore. In 1965, Moore wrote a paper observing that the number of components in a single integrated circuit doubled every two years, and that he thought this exponential growth rate would continue. Simply put, Moore's Law says that more computing power begets more computing power.[9] Or, as Asimov wrote in 1978: "Technology is a cumulative thing . . . machines are invariably improved, and the improvement is always in the direction of . . . less human control and more auto-control—and at an accelerating rate."[10]

Given that strong computer power is one of the keys to more advanced Artificial General Intelligence, I'm willing to go out on a limb with the majority opinion and predict that by the middle of this century, we could have AGI systems as intelligent (but I hope not as murderous) as HAL, and personal digital assistants as intuitive and amorous as the Scarlett Johansson–voiced character in *Her*. Our synthetic pals could become part of day-to-day life by the middle of this century, if not sooner—roughly a hundred years after the publication of *I, Robot*.

What will life be like for aging Jonesers then? How will we continue to adapt? Will we treat robots like people (and people like robots)? And, without us noticing, what's already changed in the areas of life most affected by AI and bots—the world of big data, which all of us, like it or not, enter every time we make a purchase, post a Tweet, or download an app?

A week rarely goes by without clickbait popping up on social media about a new robot that is eerily, queasily human. Robotics trade shows and business conferences feature keynote speeches by Chief Robotics Officers selling a future that includes robots in retail stores, fast-food restaurants, warehouses, and boardrooms. Soon they may be cleaning our homes, doing our laundry, making dinner for our families, teaching our kids, assessing our cognitive functions, and having sex with us. The overall impression is that it's the robots' world: we just live in it. Or rather, they live in ours: the World of Tomorrow will have to adapt to the roads, houses, and office towers of yesterday, many of which were built long before serious thought was given to the role of robots and artificial intelligence in day-to-day life. Autonomous cars will navigate country roads and narrow Roman streets. Smart appliances and the Internet of Things will be retrofitted into century homes. The final decades in the lives of Generation Jones promise to be years of transition, upheaval, and adaptation—in other words, the ever-changing world we've always known will just keep on changing.

Reason seven: Right now I have more questions than answers about robots, and I'll bet you do, too. Looking forward to speculate about the future of human-robot relationships, I've been wondering how "thinking machines" will transform private life (if such a thing will even exist anymore). What will be the new normal?

Will vulnerable people in need of care—such as frail seniors and disabled persons—become emotionally attached to their robot caregivers, and vice versa? What happens to a social robot when the human in their charge dies? Is the robot simply reprogrammed? Will they be allowed to store memories of the person they used to care for? Could they access those memories to help them learn how to comfort and care for other humans?

Will autonomous cars make us a world of passive passengers or will we be more like airline pilots, turning on the autopilot function when we want to catch a quick snooze behind the wheel? If so, will we simply become semi-competent quasi-caregivers to our own machine servants?

With nothing much for humans to do in a robot-controlled economy, and virtual reality travel available to all, why will we need to *go* anywhere, anyway?

Will our brains (not to mention our bodies) atrophy when all executive functions are looked after for us? Or will the labor of robots free us to become those pale, tall, white-robed smarty-pants philosopher societies we remember so well from *Star Trek*?

Can decision-making robots be programmed with a moral code? Is there such a thing as a good robot versus a bad one?

Let's assume that Moore's Law keeps up the momentum of advances in AI research, that machines get smarter and that we change along with them. The Jonesers who are still alive to see their hundredth birthdays may live long enough to cheat death, by simply morphing into cyborgs. (Before you write that off as crackpot dreaming, consider this: Google's biotechnology company, Calico, has been working on a solution for death since 2013.)[11] Will the unexpected consequences outweigh the benefits? Some scholars, including Stephen Hawking and the professors at Oxford University's Future Of Humanity Institute, worry that super-intelligent machines could threaten the very future of humanity and totally disrupt the world as we now know it.[12]

And yet . . .

If robots are so bad for us, why do we love stories about them so damn much?

Is it because as humans, we keep letting ourselves down?

Perhaps robots are humanity's true children, offspring who are not only guaranteed to reach their full potential but who can gallop tirelessly to keep up with the ever-accelerating, nonstop work cycle. We may have inadvertently created a society where only robots can function without breaking down.

As older adults, the question is, will our neutrally enhanced automated children—our HALs—grow up to care for us, or eliminate us? Oppress us, or free us to pursue loftier interests, free from the burden of work? Will there come a time when we must either yoke ourselves to the robots or die as a species?

On the bright side, AI could offer solutions to the all-too-human problems of aging, disability, disease, and traffic accidents. And, if Asimov is right, perhaps, one day, robot lawmakers will prevent us from destroying the world—if rogue AI machines don't destroy humanity first.

One thing is sure: some Jonesers will live long enough to witness the growing impact of AI. We can use our remaining time on Earth for good rather than evil—including thinking about how robots and humans should coexist and how to make both robots and human beings behave more humanely.

Let's look back at what a long, strange ride it's been for us, growing up as astronauts into our own future—and forward to what awaits us as we finally touch down in our long-awaited World of Tomorrow.

GENERATION ROBOT

Chapter 1

ISAAC'S KIDS

1950

Sizzling Saturn, we've got a lunatic robot on our hands!

—Isaac Asimov, *Astounding* magazine

I was born in the middle of the big, fat fifties, a decade stuffed with lardy piecrust, Fluffernutters, and fear. According to the Merriam-Webster.com Dictionary, my birth year saw the first known use of Cold War terminology like first-strike, dirty bomb, antimissile, Turing test, and postapocalyptic. With the Doomsday Clock of the Atomic Scientists set at an alarming two minutes to midnight, the world was closer to Mutual Assured Destruction, better known by its grim acronym MAD, than it would ever be again, even during the Cuban Missile Crisis.[1]

I not only picked the wrong time to be born, but the wrong place: huddled on the border between Canada and western New York State, the Niagara Peninsula may have seemed a sleepy backwater, all farms and factories and tourist traps, but as my dad pointed out, "We'll be the first to go." His favorite magazine, *Popular Science*, said that Niagara Falls was a first-strike target for the Soviets because the hydroelectric station powered America's eastern seaboard, right down to Washington, DC.

The possibility of death from above was a gray thundercloud on the robin's-egg-blue sky of my childhood, starting with the basketball-sized Soviet satellite, Sputnik. We arrived about a year apart: I, on October 15,

Kryptonite appears after the atomic age begins.

Even the Man of Steel was not immune from the nuclear threat. In 1949, *Superman* comics introduced Kryptonite, a deadly radioactive by-product of the atomic blast that destroyed the Man of Steel's home planet, Krypton.[2]

1956, Sputnik on October 4, 1957. I had barely blown out the candles on my first birthday cake when the Soviets were at it again, launching poor little Laika the dog on Sputnik II. Meanwhile, down on Earth, I slept my cozy baby sleep, my capitalist cats curled in a box, safe from being blown into orbit. But the grown-ups had bigger worries than pets in space: the Soviets had the jump on us. They not only had the H-bomb, but, with Sputnik, eyes in the sky. The metal mouths of air-raid sirens gaped from the rooftops of high buildings and a red-and-black flyer appeared in our mailbox with a checklist to help us turn our cellar into a bomb shelter: stock up on canned goods, radio batteries, and water, and get ready to ride it out. The flyer also explained how to brace yourself for a nuclear attack: crouch against a good, solid wall and put your arms over your head.

I enjoyed looking at the drawing of the nuclear family in the flyer, the little girl taking cover in her crinoline dress. I guessed that the Russians had attacked while she was on her way to a birthday party.

Every Saturday morning, high-pitched as a dentist's drill, a thirty-second tone bled into our cartoons, followed by a voice reassuring us that *this is a test, this is only a test. If this were a real emergency, you would receive instructions for the Niagara Frontier.*

We knew that the Emergency Broadcast System tone would be the last thing we'd hear before atomic light flooded our cellars and crawl-spaces, and our retinas scorched and our irradiated skin sloughed off like wet Play-Doh. One day, the alert would be real, but until then we could go back to laughing at the hapless Russian spies Boris and Natasha on *The Rocky & Bullwinkle Show*. Fear became a reflex that we exercised until it hardened into muscle memory. As cultural historian Spencer R. Weart has pointed out, "Many children understood that no matter how nicely they obeyed instructions, their chances of surviving an attack might not be very high."[3]

It seems remarkable today that kids my age didn't grow up to be chronically anxious, paranoid, and suffering from low-level PTSD. (Or, maybe we did—it could go a long way toward explaining my generation's obsessive-compulsive tendencies, from narcissism and helicopter parenting, to our habit of overmedicating and over-insuring ourselves.) In the 1970s, research into the lasting effects of nuclear fear on the psyche of an entire generation found that young adults clearly remembered and even relived their childhood fears of the bomb.[4]

"I'm sure the authorities have everything under control, Johnny!"

At the movies, monsters were dredged up out of radioactive sludge to entertain teenagers, the most famous being *Godzilla* (1954), a Japanese dinosaur-like creature blasted off the ocean floor by nuclear testing. Soon, gigantic irradiated insects, animals, and even vegetables were crowding movie screens, often starring a teenage hero who tries to warn the grown-ups about the looming threat, but is dismissed with a statement like: "The authorities have things under control, Bobby/Janie/Billy/Sally: go to your room and do your homework."

There was only one safe place: the future, also known as the World of Tomorrow, when technology would leapfrog over the nuclear threat. World leaders and their brilliant, rescued-from-Nazi-Germany rocket scientists would surely come up with clever ways to save civilization (at least, the capitalist part of it, not those populated by godless Communists). We needed spaceships to help us escape our Krypton, and intelligent machines that could think more clearly than we could, if we had any hope of survival as a species.

But what *exactly* would our future look like? And would we manage to reach it before we blew ourselves to kingdom come?

When it comes to predicting the future, we citizens of the twenty-first century are an arrogant bunch. In a world of T.E.D. Talks, and "Futurist" as a job description, we're confident that we have a handle on what our lives will look like in twenty or thirty years, unlike those knuckleheads in the 1960s who were obsessed with jet packs, moving skywalks, and flying cars. Our parents' generation was advanced enough to put a man on the moon—how could they not have seen the Internet coming?

Yet, if you look *further* back in time, to the late 1940s and early 1950s, it's remarkable how much a few influential thinkers actually got *right* about the future, predicting information management systems, intelligent automation, and machines that could think and learn. Their prescience is especially surprising when you consider the limits of day-to-day technology, circa 1950: TV was new (and there still wasn't much to watch); telephone lines were usually shared (and a trans-Atlantic call had to be booked, in advance, through an operator); a "computer" wasn't a machine but a human who did calculations; horses delivered milk, bread, and eggs; and iceboxes and wringer washers were standard household appliances. In my little Canadian border town, people were still running the British Union Jack up the flagpole instead of the now-familiar red maple leaf.

Yet two visionaries from that era anticipated the future that is unfolding around us now, more than sixty years later. Both were scientists and writers. Both would write bestselling books that captured the public's imagination in the years immediately following World War II. Both believed in a future dominated by thinking machines (i.e., computers), sentient, trainable robots and what we would now call artificial intelligence. Both invented new words to describe fields of study that didn't yet exist. But while one, Isaac Asimov, believed robots would be our saviors, protectors, and benevolent governors, the other, Norbert Wiener, believed the value of intelligent automation to humanity might be offset by its potential to dehumanize us.

Wiener seems like an unlikely candidate to write one of the bestselling books of his time. Born in 1894, a child prodigy and the son of a Harvard professor, he grew up to be a brilliant, stereotypically absent-minded mathematician who joined the faculty of the Massachusetts Institute of Technology (MIT) in 1920. He doesn't sound like the kind of guy you'd invite to a dinner party for his sparkling personality: a number of the anecdotes that crop up about Wiener center around the fact that he was an unrepentant cheapskate with a tendency to pontificate. Nevertheless, *Cybernetics: Or Control and Communication in the Animal and the Machine* went through at least five printings in 1948 alone, and more in the years that followed, despite being peppered with mind-numbing mathematical equations. *Cybernetics* anticipated a number of technical fields that didn't exist yet—information technology, computer science, and artificial intelligence among them. Wiener predicted that one day intelligent machines would not only perform repetitive tasks but also think and learn. This could be great for humanity, suggested Wiener, or catastrophic: it was completely possible that machines might take control away from humanity and render us obsolete.

Because Wiener's book was written primarily for a technical audience, no one was more surprised than Wiener himself by its popularity. One *New York Times* book reviewer enthused that it was "one of the most influential works of the twentieth century" and compared it to the work of Galileo, Rousseau, and John Stuart Mill.[5] British technological historian Jeremy Norman recently described *Cybernetics* as "a peculiar, rambling blend of popular and highly technical writing" and wryly observed that despite the vast number of copies sold, "most were probably not read in their entirety by their purchasers"[6]—perhaps not unlike copies of Hawking's *A Brief History of Time* (1988) that never saw their spines cracked.

In 1950, Wiener followed up *Cybernetics* with *The Human Use of Human Beings*, an easier read for nonscientists that included less math and more human-centered philosophy and politics. Wiener, who had been involved in the war effort as a designer of antiaircraft systems, was opposed to scientists working closely with government or the military because he saw mathematicians as potential "armorers" in future wars—a nod to the role of scientists in developing the fearsome atomic bomb.

The word Wiener invented—*cybernetics*, from the Greek word for steering or navigation—is only used in technical fields today, but *cyber* lives on as a prefix for all things digital: cyberspace, cyberpunk, cybercrime, Cyber Monday, and so forth. More importantly, *Cybernetics* explained information technology and computers (although Wiener didn't call them by that name) in a way that most Americans could understand—that is, if they actually read the book.

Just as Wiener predicted, less than a decade after the publication of *Cybernetics*, banks and other large businesses started using mainframe computers manufactured by IBM, General Electric, Univac, Honeywell, and others to process payments and store large amounts of information on punch cards and magnetic tape. Computer "brains"—the first computer chips—held out the possibility of a rapid acceleration in computing power that could in turn produce the trainable robots Wiener also had written about.

Wiener's book set off shock waves in the world of science and technology. In 1951, the co-inventor of the transistor, William Shockley, wrote to his employers at Bell Labs about his goal of building a robot that could be trained to replace human workers in a memo so prescient, it accurately described real robots that would go into development more than sixty years later.[7] In 1955, he went on to start up his own company, Shockley Semiconductor Laboratory, near his hometown in Palo Alto, California. As more and more tech companies set up shop near Shockley's headquarters, the area gained the nickname we know it by today: Silicon Valley.

Back on the East Coast at MIT, cognitive scientist Marvin Minsky had a vision similar to Shockley's. Minsky had built a "learning machine" in 1951 and believed that there was very little difference between a human brain and a computer. So great was his faith in the inevitability of higher machine intelligence that he was once famously quoted as saying that, if we were lucky, artificial intelligences might one day keep us around as pets.[8] In the 1960s, Minsky would be consulted by filmmaker Stanley Kubrick about the likelihood of computers being able to speak by 2001.[9]

Neither Shockley nor Minsky shared Wiener's reservations about the dehumanizing effect of thinking machines. For them, it was damn-the-torpedoes, full-steam-ahead into a future thick with intelligent automation. When Minsky and his friend and colleague, mathematician John McCarthy, founded their lab at MIT in 1955, they decided against calling their new field of study "cybernetics" because they considered Wiener a bit of a bore. Instead, they invented their own term: *artificial intelligence.*

Even kids were getting in on the action. By 1962, a precocious twelve-year-old New Yorker, Ray Kurzweil, had started hanging around electronics stories on Canal Street in Manhattan, gathering parts to build his own versions of IBM computers. Fifty years later, Kurzweil would become one of the world's leading inventors and futurists, with a following among Silicon Valley techno-billionaires who believe that, by 2050, artificial intelligence will surpass human intelligence and, at the same time, give us technologies that will let us live as long as we like, possibly forever.[10] (We'll meet Ray Kurzweil again in the last chapter).

But arguably the most important roboticists—at least to hundreds of thousands of North American factory workers, such as my dad—were two men who met at a New York cocktail party in 1956. Inventor George Devol and marketing entrepreneur Joseph Engelberger started chatting about Devol's latest invention, which went by the snappy name, "Programmed Article Transfer Device." Engelberger, a fan of science fiction writer Isaac Asimov, said that Devol's idea "sounded like a robot."[11]

There's a good chance that as Devol and Engelberger were sipping martinis and talking robots, Engelberger's hero, Asimov, was hunched over a typewriter in his New York apartment, inventing stories about them. A compulsive writer and homebody—possibly, an agoraphobic—Asimov hated to travel: ironically, for a writer who specialized in fantastic tales often set on distant worlds, he hadn't been in an airplane since being flown home from Hawaii by the US Army after being released from service just before a test blast of the atomic bomb on the Bikini Atoll. (Asimov once grimly observed that this stroke of luck probably saved his life by preventing him from getting leukemia, one of the side effects that afflicted many servicemen who were close to the blast.)[12]

By 1956, Asimov had completed most of the stories that cemented his reputation as the grand master of science fiction, and set the ground rules for a new field of study called "robotics," a word he made up. Unlike Engelberger, Devol, Shockley, Minsky, and other scientists and entrepreneurs of the 1950s, who were not well-known outside of the business and scientific communities, Asimov, like Wiener, was famous, his books so commercially successful that he quit his job as a tenured chemistry professor at Boston College to write full-time. Asimov's 1950 short story collection, *I, Robot,* put forward a vision of the robot as humanity's friend and protector, at a time when many humans were wondering if their own species could be trusted not to self-destruct.

Born in January 1920, or possibly October 1919—the exact date was uncertain because birth records weren't kept in the little Russian village where he came from—Asimov emigrated to Brooklyn in 1922 with his parents. Making a go of life in America turned out to be tougher than they expected, until his father scraped together enough money to buy a candy store. That decision would have a seismic impact on Isaac's future, and on robotics research and the narratives we tell ourselves about human-robot relationships to this day.

As a kid, Isaac worked long hours in the store where he became interested in two attractions that pulled in customers: a slot machine that frequently needed to be dismantled for repairs; and pulp fiction magazines featuring death rays and alien worlds. Soon after the first rocket launches in the mid-1920s, scientists announced that space travel was feasible, opening the door to exciting tales of adventure in outer space. Atomic energy—the source of the death rays—was also coming into public consciousness as a potential "super weapon." But both atomic bombs and space travel were still very much in the realm of fiction; few people actually believed they'd see either breakthrough within their lifetimes.

The genre of the stories in the pulps wasn't new. Fantastical tales inspired by science and technology went back to the publication of Mary Shelley's *Frankenstein* in 1811, which speculated about the use of a revolutionary new energy source, electricity, to reanimate life. Jules Verne, H. P. Lovecraft, H. G. Welles, and Edgar Rice Burroughs all wrote novels touching on everything from time travel, to atomic-powered vehicles, to what we now call genetic engineering. But the actual term, "science fiction," wasn't coined by any of them: that distinction goes to Hugo Gernsbeck, editor of the technical journal, *Modern Electrics,* whose name would eventually be given to the HUGO, the annual award for the best science fiction writing.[13]

Gernsbeck's interest in the genre started with a field that was still fairly new in his time: electrical engineering. Even in 1911, the nature of electricity was not fully understood, and random electrocutions were not uncommon; electricians weren't just tradesmen, but daredevils, taking their lives in their hands every time they wired a house or lit up a city street.[14] Gernsbeck, perhaps gripped by the same restless derring-do as his readers, wasn't satisfied with writing articles about induction coils. In 1911, he penned a short story set in the twenty-third century and serialized it over several issues of *Modern Electrics,* a decision that must have baffled some of the electricians who made up his subscribers. At first, Gernsbeck called his mash-up of science and fiction "scientifiction," mercifully changing that mouthful to "science fiction." He went on to publish a string of popular magazines, including *Science Wonder Stories, Wonder Stories, Science,* and *Astounding.* (Gernsbeck's rich imagination didn't stretch far enough to come up with more original titles.)

Asimov's father stocked Gernsbeck's magazines in the candy store because they sold like hotcakes, but he considered them out-and-out junk. Young Isaac was forbidden to waste time reading about things that didn't exist and never would, like space travel and atomic weapons.

Despite (or possibly because of) his father's objections, Isaac began secretly reading every pulp science fiction magazine that appeared in the store, handling each one so carefully that Asimov Senior never knew they had been opened. Isaac finally managed to convince his father that one of Gernsbeck's magazines, *Science Wonder Stories,* had educational value—after all, the word "science" was in the title, wasn't it?[15]

Isaac sold his first short story when he was still an eighteen-year-old high school student, naively showing up at the offices of *Amazing Stories* to personally deliver it to the editor, John W. Campbell. Campbell rejected the story (eventually published by a rival Gernsbeck publication, *Astounding*) but encouraged Isaac to send him more. Over time, Campbell published a slew of stories that established Isaac, while still a university student, as a handsomely paid writer of science fiction.

When you read those early stories today, Asimov's weaknesses as a writer are painfully glaring. With almost no experience of the world outside of his school, the candy store, and his Brooklyn neighborhood—and no exposure to contemporary writers like Hemingway or Fitzgerald—Isaac fell back on the flat, stereotypical characters and clichéd plots of pulp fiction. Isaac did have one big thing going for him, though: a science education.

By the early 1940s, Asimov was a graduate student in chemistry at Columbia University, as well as a member of the many science fiction fan clubs springing up all over Brooklyn whose members' obsession with the minutiae of fantastical worlds would be familiar to any ComicCon fan in a Klingon costume today. Asimov wrote stories that appealed to this newly emerging geeky readership, staying close enough to the boundaries of science to be plausible, while still instinctively understanding how to create wondrous fictional worlds.

The working relationship between Asimov and his editor, Campbell, turned into a highly profitable one for both publisher and author. But as Asimov improved his writing and tackled more complex themes, he ran into a roadblock: Campbell insisted that he would only publish human-centered stories. Aliens could appear as stock villains but humans always had to come out on top. Campbell didn't just believe that people were superior to aliens, but that some people—white Anglo-Saxons—were superior to everyone else. Still a relatively young writer and unwilling to walk away from his lucrative gig with Campbell, Asimov looked for ways to work around his editor's prejudices. The answer: write about robots. Asimov's mechanical beings were created by humans, in their own image; as sidekicks, helpers, proxies, and, eventually, replacements. Endowed with what Asimov dubbed "positronic brains," his imaginary robots were even more cleverly constructed than the slot machine in the candy store.

Never a hands-on guy himself, Asimov was nonethless interested in how mechanisms worked. Whenever the store's one-armed bandit had to be serviced, Isaac would watch the repairman open the the machine and expose its secrets. The slot machine helped him imagine the mechanical beings in his stories.

Although Asimov can be credited with kick-starting a generation's love affair with robots, he was far from their inventor. (Even *I, Robot* borrowed its title from a 1939 comic book of the same name written by a pair of brothers who called themselves Eando Binder, the name eventually bestowed on the beer-swilling, cigar-smoking robot star of the TV show, *Futurama* [1999-2013].) But in writing his very first robot story, Asimov was both jumping on a new obsession of the 1920s, and mining old, deep myths going back to ancient Jewish tales of the golem, which was a man made of mud and magically brought to life, as well as stories as diverse as Pygmalion, Pinocchio, and engineering wonders like the eighteenth century, chess-playing Mechanical Turk, and other automatons.

Robots have an ancient history and a surprisingly whimsical one. Automatons have been frog marching, spinet playing, and minuet dancing their way out of the human imagination for hundreds, if not thousands, of years, but it wasn't until the machine age of the early twentieth century that robots appeared as thinking, reasoning substitute humans.[16] The word robot—Czech for "mechanical worker"—wasn't coined in a patent office or on a technical blueprint, but as the title of a fantastical play by Karel Capek, *Rossum's Universal Robots,* which was first performed in 1920, the reputed year of Isaac Asimov's birth. In adopting robots as his main characters, and the challenges and ethics of human life in a robotic world as one of his central themes, Asimov found his voice as a writer. His robots are more sympathetic and three-dimensional than his human characters. In exploring the dynamics of human-robot partnerships—as Asimov would do particularly well in detective/robot "buddy" stories, such as his 1954 novel *Caves of Steel*—he invented a subgenre within the broader world of science fiction.

Like Asimov's humanoid robots, the Three Laws of Robotics had legs. More whimsical than scientific, they established ground rules for an imaginary world where humans and mechanical beings coexisted. Eventually, the Three Laws were quoted by researchers in two academic fields that were still unnamed in the 1940s: artificial intelligence and robotics.

The Three Laws of Robotics

First published in 1942 as part of Asimov's fourth robot story, "Runaround," the Three Laws stated that:

1. A robot may not injure a human being or, through inaction, allow a human being to come to harm.
2. A robot must obey the orders given it by human beings except where such orders would conflict with the First Law.
3. A robot must protect its own existence as long as such protection does not conflict with the First or Second Laws.[17]

According to Asimov's biographer Michael Wilson, "Asimov was flattered that he had established a set of pseudoscientific laws. Despite the fact that in the early 1940s the science of robotics was a purely fictional thing, he somehow knew that one day they would provide the foundation for a real set of laws."[18] The Three Laws would continue to appear not only in the world of robot-driven books and films—like *Aliens* (1986), where the laws are synopsized by the synthetic human Bishop when trying to reassure the robot-phobic heroine Ellen Ripley—but by some real-world roboticists and AI researchers, who are now considering how to develop a moral code for machines that may one day have to make independent, life-or-death decisions.

By the time Sputnik beeped its way over America in 1957, Asimov had stopped writing fiction to devote himself to popular science books that would explain the terrifying world Americans found themselves living in. He may also have sensed a shift in readers' tastes: a new generation of science fiction writers were publishing stories that were sexier and more subversive than Asimov's technological tales of adventure. Harlan Ellison, James Blish, and Philip K. Dick were changing the face of the genre. Asimov's friends and contemporaries Robert Heinlein and Arthur C. Clarke (creator of HAL 9000) would make the jump to the New Wave of science fiction, but Asimov thought it wiser to step away and re-create himself as a science writer, full stop. After 1957, he would not return to a fictional robot book until *The Bicentennial Man* in 1975.

The timing of Asimov's robot-driven bestsellers couldn't have been better. Asimov's robots clanked their way into pop culture as the cold war was going into the deep freeze.

Starting with *I, Robot* (largely drawn from stories previously published in pulp magazines in the 1940s), Asimov anticipated that by the middle of the twenty-first century, peaceful, rational robot lawmakers that had humanity's best interests at heart would govern Earth. Like the autonomous car that we're told is going to end the carnage of drunk and distracted driving, end traffic gridlock, and save the environment, Asimov's robots would benefit humanity by putting an end to the possibility of World War III. Perhaps Asimov subconsciously invented our modern idea of the robot as a way to save humans from our own worst impulses—violent, illogical, greedy, power-mad, weak-kneed, xenophobic creatures that we are. It was a philosophy already adopted by scientists and engineers who had worked on the Manhattan Project to build the first atomic bomb: horrified by the prospect of what they had unleashed, they urged all nations to unite under a single world government. Naively, the atomic scientists believed that the mutual assured destruction of nuclear war would provide the impetus for international cooperation. (This group of scientists would eventually create the Doomsday Clock of the Atomic Scientists, with its countdown to midnight, as a way to raise awareness.) Although one world government turned out to be an unrealized dream, the concept was explored in *I, Robot* stories about robot lawmakers who maintained peace on Earth.

Asimov's benevolent machines offered hope that—just as weapons technology was threatening our future—robotic technology could swoop in and save it. Humanity *needed* Asimov's myth of the modern robot to allow the wondrous future to rescue us from the dreadful present.

The idea of solving the threat posed by one technology with more technology (rather than by changing human behavior) has become so entrenched that we scarcely question it anymore. During a *60 Minutes* interview in 2016, journalist Charlie Rose asked Andrew Moore, former Google vice president and Dean of the School of Computer Science at Carnegie Mellon University, to respond to warnings by Stephen Hawking and Elon Musk that artificial intelligence could spell the end of humanity. Moore's response was notable for its optimism and faith in the march of scientific progress for the good of humankind: "We're going to cause disruption as things change. But when I think about the biggest problems of the world—terrorism, mass migration, climate change—I don't feel helpless. I feel this generation of computer scientists is actually building technology to put things right."[19]

We are not as far away from both the technological fear and hope of 1950 as we might like to think.

In the fifties, Walt Disney worked hard to create a mind-set in children that would see the friendly side of technology, using a combination of animation, storytelling, and education, usually with a good-looking German scientist in a suit leading the lesson.

In January 1957, the *Disneyland* television show aired a program called "Our Friend The Atom." As always, the show opened with the cartoon fairy, Tinker Bell, conjuring up one wondrous world after another.[20] "Our Friend The Atom" introduced us to a wonder of Tomorrowland.

Guided by Dr. Heinz Haber, the program opened with familiar images of mushroom clouds, then segued to an animated version of the Fable of the Fisherman and the Genie: a fisherman catches a mysterious bottle in his net which releases a malevolent genie. The fisherman, whose powers of persuasion suggest that he missed his calling as a used car salesman, manages to turn the genie from foe to friend and get him back inside his bottle; not only that, but the genie has to grant him three wishes! (Coincidentally, in *The Human Use of Human Beings*, Wiener also used the genie as a metaphor for the good and evil sides of nuclear power.[21])

Dr. Haber explained that like the clever fishermen, America's scientists occasionally dredged up something unexpected, such as the power to annihilate all life on Earth. But physicists knew how to transform atoms from foe to friend. In case the metaphor wasn't clear enough, a cartoon genie was superimposed rising out of actual footage of a mushroom cloud.

After a brief lesson in the history of nuclear physics, Dr. Haber went on to explain that we needn't *fear* nuclear energy—we just needed to learn how to *control* it. Most excitingly, considering the age of the audience, Haber promised that atomic power would help us get to outer space. It was a promise us kids gulped down with our tumblers of Tang.

You have to give Disney credit for not talking down to his audience: the atomic physics sequences were beautifully animated and patiently explained by Dr. Haber, who came off less as the mad scientist type, and more as the friendly foreign gentleman next door who liked to invite your dad over for a beer. Watching the show today—as I no doubt did in reruns in the early 1960s—feels weirdly comforting. "Our Friend The Atom" was not only broadcast on TV, but also shown in schools, giving it the stamp of scientific fact.

Unless Nikita Khrushchev was a Disney fan, no one watching "Our Friend The Atom" could have known that ten months after the program aired, the unexpected appearance of the Sputnik spy satellite would send an ominous message: *Now your friend the atom can kill you anywhere, anytime.*

Two months after Sputnik beeped its way across a quivering America, the *Disneyland* TV show featured a Tomorrowland program called "Mars and Beyond,"[22] opening with Walt Disney being introduced by a humanoid robot named Garco, and puckishly narrated by none other than Orson Welles of *The War of the Worlds* radio-hoax fame. After a funny cartoon about a killer robot from Mars—which any kid would recognize as a parody of the pulp science fiction stories of the past—a new German scientist appeared to reassure us that there was already a plan in place to build a manned craft that could escape Earth's atmosphere in a mere four months, then carry us to Mars. Disney's scientist this time, Wernher von Braun, was a former member of the Nazi party and the SS, and the inventor of Germany's V-2 rocket. Gone was the comforting, soft-voiced Dr. Haber of "Our Friend The Atom," replaced by von Braun's robotic movements and stiff-speaking style. The implication was clear: if we blew up the Earth, we were already making plans to escape to other planets. Top German scientists, programmed by their previous masters to be as intelligent as computers and as driven as robots, were on the job. The space race was the happy flip side of the long-play record of the Cold War: it was only a matter of time before we'd be able to take refuge on other planets, just like Kal-El (Superboy) emigrating to Earth from exploding Krypton.

Despite von Braun's assurances in 1957 that we had a plan to reach Mars, we're still working on it sixty years later, and for exactly the same reason: Earth is doomed. As SpaceX founder Elon Musk said in a 2013 interview: "The future of humanity is going to bifurcate in two directions: either it's going to become multiplanetary, or it's going to remain confined to one planet and eventually there's going to be an extinction event."[23] The difference between the endgame feared in the 1950s and the one we're anticipating today is our worries are environmental, as well as nuclear: old Earth has been stripped, poisoned, and slowly cooked to death, like a medieval saint.

The possibility of humans escaping to space—and the role robots might play in taking us there—were explored in one of the most iconic science fiction films of the 1950s, *The Day The Earth Stood Still* (1951). Released when wartime memories were still raw, the film carries a noir-ish, world-weary tone more in keeping with the 1940s than the '50s. The story opens with an intergalactic visitor named Klaatu arriving in Washington, DC, by flying saucer—and whenever there's a flying saucer in a 1950s movie, you can be sure there's a robot inside. Instead of the "little green man" stereotype, Klaatu (played by a charismatic Michael Rennie) was simply a man, albeit one skilled enough in higher mathematics to correct the chalkboard equations of an Einstein-like professor reputed to be the smartest person on Earth. (In post-*Cybernetics* popular culture, advanced math skills are *always* the marker of a superior civilization.) Klaatu's sidekick is a huge robot named Gort, a featureless, walking atomic weapon who appears menacing but is less aggressive than the soldiers and police who shoot and eventually kill Klaatu. (Gort uses advanced technology to bring his master back to life.)

While Gort harkens back to the scary killer robots of the pulp magazines of the 1920s, his role as keeper of the peace has Asimov's fingerprints all over it. Because Gort can disable and kill humans (at Klaatu's command), he doesn't appear to be governed by the Three Laws, except in one important respect: he has been tasked to destroy humanity for the greater good of the universe, unless the citizens of Earth wise up and stop killing one another. Because Earth has "rudimentary atomic energy" and is experimenting with rockets, Klaatu's people realize it is only a matter of time before Earthlings, who are known throughout the universe for being dangerous, violent, and stupid, escape their own planet and spread their aggression to other worlds like a contagion.

At first, Gort's power is only hinted at—until Klaatu and a young war widow, Helen Benson (Patricia Neal) find themselves stuck in an elevator after Gort temporarily shuts down the planet's electrical grid in an attempt to scare Earth onto the straight and narrow. When Klaatu explains that he's worried about what Gort might do in his absence, Helen responds: "But he's a robot—what can he do?" Klaatu's grim reply: "There's no limit to what he could do. He could destroy the world."

Klaatu's message comes down to *Love one another . . . or else.* It's Gort's job to back up those words with action: he is a member of a "race of robots," created by Klaatu's people to act as intergalactic police: "In matters of aggression, we give them absolute control over us . . . we live in peace."

With its themes of rebirth and Last Judgment-style retribution, the film could be interpreted as a metaphor for the Second Coming of Christ, if Christ was accompanied by a giant robot equipped with a death ray.

As Klaatu and Gort fly off in their flying saucer, the movie ends on an ominously ambiguous note. Will Earth survive by embracing world peace or will humans allow themselves to be destroyed? It was a question still much in the minds of moviegoers and everyone else in 1951.

In 1956, the robot as emotionless protector/destroyer was replaced by loveable Robby the Robot from the campy intergalactic soap opera, *Forbidden Planet.* Unlike Gort, Robby had a speaking voice and a sparkling personality, along with the usual robot qualities of super strength and devotion to its human masters. Unlike the faceless death ray-equipped Gort, Robby's grid-like head suggested humanoid features. The anthropomorphization of robots, so dominant in Asimov's work, became the norm: in Robby's role as a good-natured friend and protector, it's easy to draw a direct line from him (or it) to Will Robinson's Robot in the TV show *Lost in Space* (1965–1968), and *Star Wars*' R2D2 and C3PO, a little more than twenty years later.

At a time when humanity could no longer be trusted to look after themselves or their planet, for us kids, the arrival of a friendly robot meant that we were no longer alone. A smarter, stronger being would always be there for us.

While robots in pop culture of my earliest years began to establish themselves as our good buddies, scientists and engineers had begun to look at robots as either helpers for humanity or replacements for human workers. Inventors were also starting to think in practical terms about what a robot might look like and do.

The notion that a robot should look like a human, with a head, limbs, hands, and a torso, grew not only out of pop culture notions of "mechanical men," but the belief that a machine that moved through and worked in a human world had to be about the shape and size of the average human, with a similar grip, reach, and finger movement. Otherwise, how could it operate an assembly line machine or drive a car? The idea that the robot might actually *be* the car hadn't emerged. Neither had the knowledge that it was possible for a robot to look *too* human, causing an uneasy or even queasy sensation in some people—the so-called Uncanny Valley effect. Today, roboticists are developing humanoid "social robots" that respond to us in emotionally appropriate ways in order to build more trust and rapport with humans, and also to learn what we're thinking and feeling by reading our facial expressions. The goal is to create human-robot relationships that mirror human ones, a necessity if robots are to take on some of the roles anticipated for them, such as caregivers, teachers, and social workers.[24]

Early on, two divergent attitudes toward human-robot relationships began to evolve—that of robot as assistant for humans (like Robby) versus robot as substitute human (like Gort, the intergalactic cop). Engelberger, the martini sipping, science fiction–loving entrepreneur, was in the Robby camp. But unlike Robby, Engelberger's first robot would simulate only one part of the human body: the arm.

Within a year of his chat with Devol at the cocktail party, Engelberger had found the capital to finance Devol's invention. By 1959, they had a prototype called UNIMATE 001. With Asimov's Three Laws of Robotics in mind—the first and most important being not to harm humans—Engelberger focused on building a robot that would take over the most dangerous tasks faced by factory workers, leaving safer, more rewarding jobs for humans. By 1961, UNIMATE 1900 had been patented and became the first mass-produced robotic arm for factory use, making its debut at a General Motors plant in Trenton, New Jersey.

Seven years later, UNIMATE's sleek pneumatic arm would reach north from New Jersey into a factory in the Canadian border town where I grew up, and touch my father like the hand of a mechanical god.

Chapter 2

MONOLITH

1968

> I am putting myself to the fullest possible use, which is all I think that any conscious entity can ever hope to do.
>
> HAL 9000, *2001: A Space Odyssey*

My father often came home from work with blood on his clothes. Sometimes, his own blood. Usually, somebody else's. Every day, he walked among machines that ate men alive. A dangling tie could strangle an engineer bending over to inspect a machine. A loose cuff caught in a press could lead to the amputation of a worker's hand. No big surprise that Dad only wore clip-on bow ties, short-sleeved shirts, and leather watch straps thin enough to snap under pressure. His wedding ring had been tucked away in a dresser drawer since his honeymoon. He'd seen too many guys dismembered, scalped, blinded, and electrocuted by being snagged by their clothing, jewelry, and hair.

Safety on the Job Begins Here read a sign at the parking lot entrance of the plant where he worked. Despite the warning, factories were dangerous places, in part because manufacturers considered safety measures more costly than replacing dead and injured workers. Until Richard Nixon signed the Occupational Health and Safety Act into law in 1970, fourteen thousand American workers died and another two million were disabled by workplace hazards *every single year* of the 1960s.[1]

—

An electrician and a natural problem solver—these days he'd probably have his own YouTube channel called "The Machine Whisperer"—Dad had been promoted to his employer's engineering department to keep the interconnected lines of machinery and their human operators in perpetual motion. He was the go-to guy to fix breakdowns, day or night.

One of his unpleasant responsibilities was to get the line up and running in the aftermath of workplace accidents that might cause a machine to malfunction (or had been caused by a malfunctioning machine).

I could tell when someone had been mauled by one of the mechanical monsters by the look on Dad's face when he came home for lunch, a legacy of my family's Italian tradition of gathering for a big midday meal. On bad days, he'd sit at the table silently, his starched white shirt stained with blood and grease. We could see him struggling not to unburden himself on us, until he'd finally mutter something like: "Hell of a thing happened this morning."

Mom, anticipating what he might say next, would warn: "Not while we're eating," and ladle more *penne arabbiata* onto his plate. My grandmother, Nonna Tina—Dad's mother—would stare at him uncomprehendingly from her wheelchair at the opposite end of the table.

"What happened, Dad?" As the youngest and last kid at the lunch table, I was the only one who wanted to hear all the gory details.

By 1968, Dad was censoring himself less often. He would run his hand over his balding head, and talk. An aluminum ladder, against an iron catwalk, and a spark: boom, a man was electrocuted to death. A worker's arm caught by his sleeve. Boom, arm gone. A shard of hot metal flying straight into a machine operator's eye. Dad suspected he'd never see out of it again.

"Tee," my mom would warn, using the nickname he had picked up in school because no one could pronounce his Italian name, Attilio.

My father—usually chatty, funny, and affectionate—would press his lips together, stare at his plate, and swallow the horror of what he'd seen along with his lunch. Afterwards, he'd change his bloody shirt and go back to work.

Given the sometimes-grisly nature of Dad's job, his enthusiasm was understandable when he came home for lunch one day and told us that he had been put in charge of UNIMATE, the world's first factory robot, manufactured by Unimation of Danbury, Connecticut.

My mom made a face. "Robot? *Ma che,* it'll steal jobs from the other guys in the neighborhood."

"It'll save lives," insisted Dad. He explained that UNIMATE had been programmed to do one particularly dangerous job: lift a massively heavy chunk of hot metal out of a mold, dunk it into coolant, and put in a press to form the steering column of a car, over and over again. Sounds simple, but it was hazardous work for a man doing it seven hours a day, fatigue being a big factor in accidents—not to mention the intense heat, falling ash, and the never-ending, earsplitting din of machinery going full tilt.

Rise of the intuitive engineer

Star Trek's Montgomery "Scotty" Scott (James Doohan in the 1968 original series) possessed a deeply intuitive streak that suggested he had a psychic connection to the ship's engines, which in turn made the USS *Enterprise* itself seem "human." The same suggestion of an emotional human-machine relationship appeared in the cult TV series *Firefly* (2002–2003), when ship engineer and flower child Kaylee Frye (Jewel Staite) stroked the interior hull of the ship *Serenity* after a taxing getaway, while murmuring, "That's my girl. That's my good girl."

Assigning this one nasty, dirty, tiring task to a robot meant that a human—prone to exhaustion and inattention—didn't have to risk life or limbs doing it. Besides, Dad reasoned, the workers replaced by UNIMATE would get new jobs programming, building, and maintaining robots, work that would become increasingly important after the moon landing, when NASA started building lunar colonies for settlers.

"It only stands to reason that we'll need robots to build geodesic domes so people can live on the moon, hon," said Dad, motioning for a second helping of polenta with anchovies, or whatever Mom happened to be serving that day.

"*Ma che,* lunar colonies," muttered Mom darkly, which was her usual response to my father's crackpot ideas about living on the moon.

I glanced at Nonna, hunched over her meal in silence. Robots did not compute for her. Neither did space travel. She watched the Apollo missions on TV with a disbelief verging on hostility. It was all just play-acting, she insisted. And she couldn't remember things the way she used to: she was even starting to forget how to speak English. "Senility," the doctor called it. She'd broken her hip on a fall down the cellar stairs carrying an armful of laundry, and despite my father's repeated attempts to get her to walk again, she was stuck in a wheelchair where she billowed into a big Buddah of a woman, too heavy for my mom or even my dad to lift. If there had ever been a fire or nuclear attack, there was no way to get Nonna out of the house. She was marooned on the main floor, rolling between four rooms: her bedroom, the laundry room that Dad had retrofitted as a handicapped bathroom, the kitchen, and the rarely used living room. She could sit outside on the back stoop where Mom hung laundry but couldn't get herself to ground level to our yard, or the garden, or the vineyard beyond it.

Dad's robot might be able to save men's lives but it couldn't do anything to help Nonna, I thought, underestimating my father's willingness to solve problems with whatever he had at hand.

Dad was an unlikely candidate to be put in charge of a space-age innovation like UNIMATE. Born in a non-electrified stone cottage in a windswept corner of the Italian Alps in 1920 (the year the word robot was coined by playwright Karel Capek), Dad received a haphazard, Depression-era education after his family immigrated to Canada. He studied to be an electrician by mail through International Correspondence Schools (ICS) of Scranton, Pennsylvania, famous for advertising their technical courses in the back pages of comic books and magazines.

MIT, it wasn't. ICS sent Dad a set of leather-bound textbooks published in 1917 called *Hawkins Electrical Guide: A Progressive Course of Study for Engineers, Electricians, and Those Desiring A Working Knowledge of Electricity and Its Applications.* Already out-of-date by the time Dad started studying with ICS in 1939, *Hawkins Electrical Guide* had strangely religious overtones mixed in with lessons about semiconductors and alternating current motors. Volume One featured a glowing illustration of a saintly Thomas Edison (Dad's hero) and called electricity "one of the secrets of creation" left on earth for humankind to seek out, a claim it backed up with scripture: "It is the glory of God to conceal a thing: But the glory of Kings (i.e., wise men), to search out a matter."[2]

With little higher education than the pious *Hawkins Electrical Guide,* Dad started work at eighteen, hand-cranking open the sluice gates at an industrial mill that was still using nineteenth-century waterpower technology in the 1940s. Eventually, his problem-solving abilities brought him to the attention of Cleveland-based Thompson Ramo Wooldridge Inc., better known as TRW, a major player in the military-industrial complex. He joined the company's Niagara branch plant, Thompson Products, in 1947, as an electrician, and was promoted to the engineering department in 1962—a management job, even though he didn't have the education.

In 1968, he was a dozen years away from retiring on a company pension. It was a time of transition when the sleek silver silent UNIMATE manufactured steering columns, night and day, while Dad looked on admiringly. UNIMATE opened a fascinating new frontier of knowledge for Dad, not unlike the giant black slabs known as monoliths in *2001: A Space Odyssey,* which had shown humankind how to do everything from use primitive tools to fly to the stars. It was the must-see movie of 1968, the year UNIMATE descended into Dad's life. As the robot's guardian, he had one last chance to indulge his curiosity in the new world of automation and computerization that was just beginning to impose itself on the iron pistons and presses that drove manufacturing.

As *2001*'s director, Stanley Kubrick, said to science fiction novelist and inventor of the communication satellite, Arthur C. Clarke: "What I want is a theme of mythic grandeur."[3] They started collaborating in 1964, writing the screenplay and novel simultaneously under the working titles, *How the Solar System Was Won* and, later, *Journey Beyond The Stars*, before finally settling on *2001: A Space Odyssey*.[4] Kubrick was just coming off the success of his 1964 black comedy *Dr. Strangelove*, a movie that reflected his belief that a nuclear world war was inevitable. He approached Clarke, not only because he wanted his help in making a "really good" science fiction film, but because he wanted to present a hopeful vision about the future of humanity—namely, that with the help of extraterrestrial beings, we could evolve to the point where we would realize the futility of war and choose to destroy all nuclear bombs, as the Star Child does in both the book and movie version of *2001*—a point not completely clear to anyone who only saw the film.[5]

Some of Clarke's early notes about the collaboration, including a conversation about robot biology with Isaac Asimov, hint at an early vision of the movie that sounds more steampunk than sci-fi:

"October 17. Stanley has invented the wild idea of . . . robots who create a Victorian environment to put our heroes at ease.

November 28. Phoned Isaac Asimov to discuss the biochemistry of turning vegetarians into carnivores."[6]

Kubrick used every device available to filmmakers of his day to awe his audience, from the futuristic (yet rooted in the swinging sixties) sets and costumes, to the horns and kettledrums of *Thus Spoke Zarathustra*, to the oh-wow-man psychedelic special effects that won the film its only Academy Award. Bookending the central story of a trip to Jupiter (Saturn, in the novel) were the monoliths that endowed man-apes with the intelligence to learn how to use tools, and transform astronaut David Bowman (Keir Dullea) into a "pure intelligence" known as the

Star Child—all thanks to an ancient civilization that had figured out how to preserve their higher intelligence inside robotic machines. (I like to think that caught the attention of eighteen-year-old Ray Kurzweil, future author of *The Age of the Spiritual Machine, The Singularity is Near,* and other books about achieving immortality by transferring our consciousness into robots. More about Ray is coming in the chapter "Sex and the Singularity.")

Kubrick created a movie that was mind-blowingly innovative on every level. Maybe a little *too* innovative: many moviegoers wondered what the heck it was all about; they'd have to read the novel, published after the movie was released, to get the gist. So quickly did *2001: A Space Odyssey* influence popular culture that when Apollo 8 orbited the moon on Christmas Eve 1968 (a "test drive" my father called it), the astronauts joked among themselves about whether to report the appearance of a large black monolith on the surface of the moon to Mission Control.[7]

Watching *2001: A Space Odyssey* today, the voyage to Jupiter is more like a nostalgia trip back to

Is HAL really IBM?

One story about HAL's name was that it was code for IBM: if you alphabetically shift forward each letter in "HAL," you'll see why. When HAL regresses, it sings *Daisy Bell* (sometimes known as *A Bicycle Built for Two*), the same tune sung by an IBM 7094 computer in 1961. In our era of blabbermouth digital assistants like Siri, it's hard to appreciate how shocked people were by the idea that a computer could talk, let alone sing.

Clarke denied the IBM connection, writing that HAL stood for: "*H*euristically programmed *AL*gorithmic computer."[8] Clarke created this origin story for HAL (or Hal, as he's called in the book):

> Artificial brains could be grown by a process strikingly analogous to the development of a human brain . . . Whatever way it worked, the final result was a machine intelligence that could reproduce—some philosophers still preferred to use the word "mimic"—most of the activities of the human brain, and with far greater speed and reliability . . . Whether Hal could actually think was a question which had been settled by the British mathematician Alan Turing back in the 1940s. Turing had pointed out that, if one could carry out a prolonged conversation with a machine . . . without being able to distinguish its replies and those that a man might give, then the machine *was* thinking, by any sensible definition of the word. Hal could pass the Turing test with ease.[9]

the 1960s' overoptimism about space travel: even Elon Musk doesn't think we'll reach Mars until after 2020. The psychedelic effects look like a stoner's black light poster, and the bombastic theme music has been parodied to the point of ridiculousness. But the movie still resonates, thanks to the durable influence of the soft-spoken, chess-playing, astronaut-murdering artificial intelligence, HAL 9000, voiced by actor Douglas Rain. If ever a fictional character embodied our conflicting feelings about robots, HAL was it.

Dad's robot was no HAL, of course, and much of the technology behind UNIMATE wasn't new: all the working parts of its two-ton pneumatic arm already would have been familiar to Dad. The magic was in the programming that made UNIMATE perform tasks without human intervention, using a drum memory contained in a base about the size of a window air conditioner—this, in an era when UNIVAC and IBM mainframe computers could fill a large room. The robot could be programmed to do two hundred different movements, including welding, gripping, drilling, and spraying, with a precision and speed far beyond the capabilities of humans. What's more, it could work sixteen hours straight without a coffee break.[10]

Sleek as a V-2 rocket, its silver finish and silent operation contrasted with the deafeningly loud, greasy machinery that surrounded it, not to mention the dirty, sweaty, fatigued, and deafened men. Like the intelligence-conferring monoliths in *2001: A Space Odyssey*, UNIMATE landed in an industrial setting that, whether the workers understood it or not, would be obsolete within their lifetime.

Dad must have suspected that they were at the dawn of a robotics revolution that would disrupt manufacturing. What he *didn't* know was the serendipity that had brought UNIMATE into existence in 1956, the same year he and Mom conceived me. George Devol and Joseph Engelberger just happened to meet and hit it off at that Manhattan cocktail party. George had an idea, Joseph had a vision. Martinis were shaken (or maybe stirred) and "I love you man!" promises were made, but unlike most gin-fueled bromances, this one survived Engelberger's morning-after hangover. Despite being turned down by forty-seven shortsighted investors, who clearly didn't appreciate the good karma automatically endowed to a project kicked off over a few friendly drinks, Engelberger drummed up the capital to turn Devol's original design into a working prototype within a year of that fateful party and the Unimation company was formed.

As mentioned in the previous chapter, UNIMATE debuted at a GM plant in 1961, but most other manufacturers were skeptical about something as wing dingy as robots. To win them over, the ever-tenacious Engelberger wrangled a guest spot for UNIMATE on the *Tonight Show*, where his automated baby showed off by pouring beer into a glass, conducting the NBC orchestra, and putting golf balls.[11] The PR stunt did the trick: who could resist a robot that that could charm Johnny Carson?

UNIMATE began appearing in factories across the United States. Just one year before the moon landing, it finally found its way to the rust-belt border town where I grew up.

Dad gave UNIMATE a not-very-original nickname, "Robbie"—not for *Forbidden Planet*'s Robby the Robot, but for "Roberta." Like a car, ship, or any other beautiful machine, in Dad's eyes the robot was feminine, its inner workings as mysterious, intriguing, and worthy of exploration as a woman's body. This sentiment was shared by Arthur C. Clarke, who wrote in the novel version of *2001: A Space Odyssey*: "Spacepods . . . were usually christened with feminine names, perhaps in recognition of the fact that their personalities were sometimes slightly unpredictable. *Discovery*'s trio were Anna, Betty, and Clara."[12]

Sure enough, Robbie surprised everyone by occasionally going haywire. As a preprogrammed robot, she could do *one* action, *repeatedly*. Pick a part out of a mold, pivot, dunk. Problems arose when she confronted anything even slightly unusual: for example, a variance in heat might make a piece of metal more difficult to remove from its mold.

Whenever this happened, Robbie interrupted her smooth flow of pick, pivot, dunk, to haul back with her shiny 4,000-pound. arm and hurl a blisteringly hot piece of metal across the shop floor, narrowly missing the heads of assembly-line workers. Dad suddenly found himself playing Astronaut Bowman to Robbie's HAL.

His solution was cheap, simple, and effective: since the robot was pitching parts, he'd surround her with baseball nets with which to catch them. It was my earliest example of what can happen when a new technology meets a problem solver with a creative imagination.

Robbie captured my imagination, too. The way Dad talked about her made me think that she looked like one of the robots I was used to seeing on TV: a whirring, mechanical biped, like the B-9 Environmental Control Robot on *Lost in Space*, better known simply as "Robot." It was exciting to imagine such a creature working alongside Dad at the plant, which I only ever glimpsed at the Thompson Products Old Guard Employee Association Christmas Party every year, a fun-filled, glittery, gift-laden orgy of sweet treats and wrapped toys, presented to each employee's child according to their gender and age. The outside of the factory was decorated as magnificently as Disney's Magic Kingdom (I assumed, never having been there), with Santa's sleigh and team of flying reindeer illuminating the sky over Plant 1, where the management offices and cafeteria were located. It was a vision of everything the sixties was about. Progress. Optimism. Never-ending wonders. Moon shots. Colored lights. And brand-name toys for all the kids. My first and only Barbie doll came to me, compliments of Thompson Products, inside a shiny black and pink carrying case with a silver latch that opened into a preteen's dream bedroom with Barbie's mod outfits on hangers and several pairs of tiny, high-heeled shoes. I almost cried with joy the year I got that gift. My parents didn't approve of teenage dolls with bullet breasts and platinum blonde hair, or ones that ate, walked, talked, or did anything robotic. Under our tree at home, I would only ever unwrap cuddly infants in frilly layettes that looked as if they had come straight from the Old Country. By contrast, the gifts I received at

Thompson Products Christmas parties were with-it and modern: whoever was buying toys on behalf of the Employees' Association was as good as Amazon's "If you like this, you'll also like" algorithm.

Every year, at some point during the party, we were all made to troop through Plant 2, the factory floor. I don't know why they led a straggling line of kids in their best party clothes, high on sugar and the off-gassing of polymers from new plastic toys, through what could be described as William Blake's vision of industrial hell: an acre of deafening machines, grimy with ash, going full tilt on a vast expanse of oil-soaked gray concrete. During our short walkabout, we kids saw and heard and smelled the iron wheels turning the machinery of our lives. It was like walking into the *Wizard of Oz* and seeing the glittering curtain pulled aside to reveal the old swindler hidden behind it, working knobs and yanking levers to keep the Great Oz alive, a giant mechanism that put food on our tables, Buster Brown shoes on our feet, and twenty-five pound turkeys in our ovens every Christmas. Night-shift workers punched out and day-shift workers punched in, but the din of the machinery never stopped: except for a two-week summer shutdown for plant maintenance, Thompson Products turned out brake linings for Ford station wagons and sedans, twenty-four hours a day. At least, that was its business in my hometown; the parent company TRW had manufactured America's very first Intercontinental Ballistic Missile and now—excitingly!—was helping build NASA's lunar lander. Although it was a giant leap from car brakes to the moon landing, we all felt like a part of it. But the shop floor didn't look anything like Mission Control in Houston, with its crew-cutted, pocket-protected technicians in button-down shirts. The earsplitting decibel level of the machinery explained why my father had such a hard time understanding us sometimes. It would still be years before ear protection was required in the plant. In the sixties, the workers simply went deaf. As we kids walked through Plant 2, we clamped our hands over our ears against the noise and screamed in pain. No one noticed because the sound of the machinery drowned us out.

I went to those parties, year after year, until I turned twelve. My very last gift was a miniature luggage set—an overnight bag, makeup case, and suitcase—as if Thompson Products were wishing me bon voyage into teenage-hood. That also happened to be the year that Robbie came to the shop floor, and Dad started puttering around with robotics to help solve problems he faced at home.

Job one was to build an automated device that could move my wheelchair-bound grandmother out of the house. As the caregiver for Robbie the rebellious robot, Dad could easily order a few spare parts from Unimation's head office in Connecticut and smuggle them home from work.

All it took was a doctored purchase order, or two.

Dad's UNIMATE-inspired creations were not his earliest automated inventions. As a teenager, he had built a device that practiced the violin for him and designed a homemade self-timer for his box camera so that he could snap photos of himself from across the room.

And then, there was his automated wine-bottling system. My grandfather died unexpectedly in 1963, leaving behind three years' worth of homemade wine in huge barrels. The wine cellar was cramped, windowless, and so low ceilinged that Dad couldn't stand without hitting his head. Nonno was in the habit of just opening the spigot and filling a bottle or two as needed, but Dad wanted to empty the gigantic barrels all at once so he could move my widowed grandmother to our house next door and rent the old place. Inspired by the assembly lines he'd seen at work, he rigged up a series of hoses that snaked their way up the cellar stairs and out the front door of Nonno's house, across the lawn, and into the basement window of our house next door, where my ten-year-old brother Ricky stood waiting to fill bottles with wine pumped through a hose hanging through the basement window. But the pump worked too fast for Ricky to keep up, turning it into a scene from the *I Love Lucy* episode when Lucy goes to work on a fast-moving assembly line. My mom found Ricky, up to his ankles in wine, drunk on fumes that had built up in our poorly ventilated basement. Mom stomped next door and Dad's dreams of automation went into suspended animation for a few years. He would not try again until Robbie inspired him to dream big once again.

We knew something was up when Dad told us to keep our mouths shut about the thing he brought home from work. A solenoid. A top-secret switch inside Robbie that controlled her movements: Dad had ordered a couple of extras in case she broke down. One of the extra solenoids found its way home from work with him.

Whistling "Red Sails in the Sunset" (his favorite work tune), he took the solenoid out back with his toolbox, a motor, and something that looked like a giant screw. Oh, and a section of ornate metal railing, leftover from a long-ago project when he'd fixed the front veranda. He built—how can I put this? An elevator on the outside of the back our house. It traveled about ten feet, from the back laundry stoop to the ground. But this was not a simple freight elevator that you operated by hand. Dad's elevator may have been open to the elements but it

Rosey, Mac, and Uniblab of *The Jetsons* (1962 to 1963)

I don't think Dad ever saw *2001: A Space Odyssey* and I doubt he read the novel. (I suspect he *did* read Norbert Wiener's *Cybernetics*, a copy of which I found in my parents' library after Dad's death.) But like the rest of us, he never missed *The Jetsons*. Even though the Jetson home was heavily automated—including a virtual reality exercise screen that did George Jetson's calisthenics for him—robot maid Rosey vacuumed, dusted, washed, and cracked wise in a Brooklyn accent. Today, Rosie could easily find work as a social robot—someone to gossip with Jane, look after Judy and Elroy, and pour a drink for George after his grueling two-hour workday at Spacely Sprockets. Rosey even fell in love with the caretaker's robot, Mac, causing both of them to temporarily malfunction.[13]

Then there was the day that George was passed over for a promotion in favor of the robot Uniblab, which badgered people to "work work work work," fired them by dumping them into the trash, and tricked George into feeding money into its built-in slot machine.[14] (Shades of the one-armed bandit at Isaac Asimov's father's candy store!) Uniblab's name may sound like UNIMATE but it's more likely a play on UNIVAC (Universal Automatic Computer), the system used by large corporations in the early 1960s to store and manage information. George's situation is one we're seeing more and more often, as artificial intelligence moves from the factory floor into the management suite, doctor's office, investment firms, ad agencies, and news services.

still offered the convenience of push-button technology. All Nonna had to do was wheel herself onto the deck of the lift, push the bottom button, and the elevator would slowly lower itself to the ground. Press the top button to reverse the process. To be absolutely sure no one fell out, the elevator would only run when a safety gate, fashioned from the decorative veranda railing, was locked in place. Ricky and I took an Apollo 8–style test drive, complete with countdown, as we wrestled to be the first one to press the button.

Once Dad was sure the elevator could bear the weight of two squabbling adolescents, Nonna was wheeled out. As Dad explained the operation of the device to her in their *Piemontese* dialect, she glared at the buttons with suspicion.

In fact, Nonna already had an escape route from the house: the year she moved in with us, Dad had built a steep wooden ramp from the stoop into the yard. The problem was, Nonna couldn't get herself down the ramp without picking up speed and losing control, and the only person in the house strong enough to keep the heavy wheelchair under control while walking down the ramp was Dad himself—or Ricky, when he was older. Dad wanted a robotic solution that Nonna could operate on her own, as easily as pushing a button.

Think of an elevator as a robot that carries you in its tummy, or a self-driving car that moves vertically. It is a form of automated transportation that's been around for so long, we no longer give it much thought, but we wouldn't have skyscrapers today without elevators. When push-button models started replacing elevator operators in the 1920s, people were as skeptical about riding in them as some of us are about self-driving cars. *How, exactly, did the elevator know where to stop?* Push-button technology seemed, at the time, to suggest some type of machine consciousness.[15] You simply pressed a number and the machine quietly, flawlessly whisked you there without being cranked to a stop by a human operator. Amazing!

Nonna rarely used the elevator, preferring to sit atop it motionless in her wheelchair, gazing out at a world she no longer recognized. It would be decades before anyone thought about robot caregivers for elderly people living with dementia, or cyborg legs to help disabled people walk again.

Ricky and I (and our friends, cousins, and various pets) rode the Nonna-Mover up and down constantly; I was still young enough to pretend it was taking me to one of the moon colonies where I was sure I was destined to live one day.

One day, lazily riding the Nonna-Mover up and down, I noticed Dad out in the middle of the yard with a large wooden spool, a huge length of metal cable, and our ride-on lawnmower. Since Dad's responsibilities at the plant made it difficult to look after the acre or so of vineyard he'd inherited from my grandfather, he decided to pull out all but a few grape rows, leaving just enough to supply himself with homemade wine, and planted grass on the empty land, resulting in a yard about the size of a football field.

It didn't take long for Dad to realize the drawback of turning a small vineyard into a gigantic lawn. You had to mow the damn thing all the time. Ride-on mowers in those days were tetchy things; the exposed motor overheated quickly and burned bare skin, something that happened to me more than once when I was taking my turn at the controls. Not to mention it easily bogged down in patches of long grass or on uneven ground.

Outer space botanist goes rogue in *Silent Running* (1972)

This cult classic bridges the imaginative gap between the powerful, malevolent HAL and the friendly bots of *Star Wars*. Freeman Lowell (Bruce Dern) is one of a skeleton crew managing a cluster of solar-powered garden domes floating near Jupiter. When he is ordered to blow up the domes containing his beloved plants and return to Earth, he instead murders his knuckleheaded astronaut companions and reprograms three worker drones, nicknamed Huey, Dewey, and Louie, to play poker and help him tend the gardens. The squat, waddling worker drones were played by three double-amputee actors in robot suits, walking on their hands.

Dad lost patience with the mower. He hated the inefficiency of riding around for hours just to keep the grass trimmed. And that wasn't the only thing he had to tend: we also had a substantial vegetable garden, a fruit orchard, and a *topia*—a fifty-foot long latticework walkway, covered with grapes, shading a concrete sidewalk that started at the back door of Nonno's old house and ran halfway across our giant lawn, where it came to an abrupt stop. The mysterious sidewalk-to-nowhere had been built by Nonno as a gift to Nonna: a cloister, through which she could walk from her house to the church two doors over, without her feet ever touching the earth or the sun burning her fair, northern Italian complexion. When she broke her hip in that fall down the stairs, and it became obvious she'd never walk to Mass, or anywhere else, again, Nonno sadly abandoned the project, leaving behind a shady fairy-tale pathway, beyond which rolled our ridiculously huge, green yard. The restless compulsion to dream and to build must have been genetic.

Dad's goal was to figure out a way to make the ride-on mower cut the grass without a driver. He started with a simple concept: attach the metal cable to the steering mechanism, jam it in place so that it went in perpetual circles, and let the mower wind itself up until it had cut the equivalent of a UFO crop circle on one side of the lawn, then move the contraption to the other side of the lawn and let it cut another huge circle. All that would be left for Dad to do was tidy some edges and in-between spots.

Watching Dad go through trial-and-error experiments with the mower was endlessly fascinating. For days, the mower would wander off in the wrong direction dragging the spool behind it as Dad jogged to the rescue, or the cable would go too slack to control the mower. From atop his impossible elevator, I figured that this was one automated idea that wasn't going anywhere but into the toolshed.

I was wrong. Dad, somehow, figured out how to set the controls just-so to keep the mower on a taut line, winding itself up like a giant yo-yo in reverse. School kids lined the wire fence behind our yard, watching openmouthed as the mower cut the lawn all by itself while Dad looked on nonchalantly.

This was only step one. He had picked up an even better idea from another new technology at work: sensors.

One of the many dangers at Thompson Products was the fleet of forklift trucks used to move just about everything on the factory floor. They were loud and usually driven too quickly in the crowded space. It didn't help that workers were either deaf or couldn't hear over the sound of machines and jackhammers that were used indoors as well as out. The resulting bings, bangs, dents, and crushed metal (and human bodies) were starting to get expensive to replace . . . and with the costly UNIMATE now exposed to a possible collision, the engineering department needed a way to both visually and audibly signal to forklift drivers that they were approaching objects or men at work. The answer was to sink concrete posts into the shop floor in strategic areas and attach sensors to them. Ricky had a summer job digging the postholes, during which time he saw UNIMATE up close; years later, he poetically described her to me as "a princess surrounded by a lot of ugly ducklings."

Dad's plan was to obtain a number of sensors and bury them underground in our yard in a grid pattern, creating a digital track for the mower to follow back and forth across the yard—an idea similar to the robot delivery systems used now in factories and hospitals. It is also the way that robotic lawnmowers work today. Sadly, despite the logic of the design he saw in his mind's eye, Dad lacked the equipment to turn it into reality. Maybe he thought that the disappearance of an acre's worth of sensors from Thompson Products would be noticed. In the meantime, his clockwork version of a robo-mower worked just fine, until he got tired of it and put in a pool.

As the 1970s drew to a close, I was away at university on a Thompson Products Scholarship, so I only heard about Dad's plans to automate the kitchen secondhand from Mom. The idea had come to him at a local jewelry store while picking out a gift for me from a display of bracelet charms. The charms were laid out on a tiny rotating Ferris wheel inside a case: you pressed a button and the wheel turned.

It occurred to Dad that by sizing up the rotating wheel, and installing it behind the walls of our kitchen, Mom could fetch any item stored in our cold cellar—where we kept a winter's worth of preserved fruits, juices, and tomato sauce—simply by pressing a button. It would be a very "Jetsons" experience: press a button marked TOMATO SAUCE and it would rise from the cellar to the kitchen cupboard. No more running up and down stairs.

By then, Nonna was almost one hundred years old and living in a nursing home. Outside of family gatherings, Mom was only cooking for herself and Dad. She told Dad that she really didn't need much help anymore. I suspect she also wasn't keen on seeing her kitchen reduced to rubble.

Reminiscing with my brother over Skype about this period of Dad's life, I asked, "Why did he want to automate everything? What was motivating him?"

On the screen of my Mac, I watched him mull over my question.

"Dad had an innovative mind. He enjoyed the creative thought process, applying whatever technology was available to him at the time. And maybe he was just trying to make life more . . ."

My brother paused again, searching for the right adjective and finally came up with: "Interesting."

Meanwhile, just as Dad was hitting his stride as the inventor of the windup robo-mower and Nonna-Mover, a robot 2,600 miles from Niagara Falls was shaking and shuddering its way around a lab in Silicon Valley's Stanford Research Institute, now known as SRI International. Named "Shakey" (because it was), it was the earliest robot to be programmed to see, move, and solve problems on its own. After three years in development, Shakey's inventors, led by AI pioneer Charlie Rosen, decided it was time for their robot to meet the world. In 1969, they shot a video with the catchy title, *Shakey: An Experiment In Robot Planning and Learning*. Backed by Dave Brubek's jazz masterpiece *Take Five*, and livened up by Charlie Rosen in a vampire cape occasionally swooping in as a box-shifting "gremlin" to confuse the robot, the video demonstrated how Shakey recognized new obstacles in its path and worked out ways to move and avoid them. Thanks to the cool background music, Rosen's Nosferatu-like troublemaker, and Shakey's mysterious beeping, the video actually managed to make twenty-eight minutes of box moving seem kind of exciting.

A teletype machine sent commands to Shakey in plain English, such as "GoTo" or "GoThru," which were in turn translated into a calculus-based planning program called STRIPS that Shakey used not only to solve problems, but to remember solutions, adding them to what the voice-over called its "action repertoire." In describing what the robot was doing, Shakey's developers used words that describe human emotional states: uncertainty, realization, remembering, predicting.

Shakey moved through doors and in and out of rooms guided by its vision program (TV cameras and preprogrammed information about its environment), and sensors called "cat whiskers" helped it to recognize and move boxes, like a giant mechanical version of Maru the box-loving Internet cat. Shakey calculated its route, learned from its mistakes, and stored what it had learned for future reference, all while tethered to its "brain," a wall-sized, 1.35 megabyte computer.

The SRI video didn't make clear how agonizingly slowly Shakey moved: it often stood immobile for many minutes, working out what to do next. But the robot was intriguing enough to catch the attention of *The New York Times, National Geographic,* and *Life* magazine. A 1970 *Life* article wildly exaggerated the robot's abilities: describing Shakey on the moon (those lunar colonies again!), the writer predicted that it could "without a single beep of direction from earth, gather rocks, drill cores, make surveys, and photographs, and even decide to lay plank bridges over crevices he had made up his mind to cross."[16]

Shakey was more than an experiment in early artificial intelligence: the government had funded the project in the hope of developing a robot that could work as a military sentry, and possibly explore space. By the time the experiment ended in 1972, it was obvious that the challenge of building a mobile robot with independent decision-making capabilities would take far longer than the developers had anticipated or the media reported.

Technology journalist John Markoff wrote that the *Life* article "hyped the machine as something far more than it actually was . . . The SRI researchers . . . were dismayed by a description claiming that Shakey was able to roll freely through the research laboratory's hallways at a faster-than-human-walking clip, pausing only to peer in doorways while it reasoned in a humanlike way about the world around it . . . the description was particularly galling because the robot had not even been operational [on the day *Life* journalist Bernard Darrach visited]."[17]

Shakey's overblown press coverage was an early example of robots failing to live up to their hype, which helped usher in an era in the 1980s known as the "AI winter," when venture capital, unimpressed with the slow pace of AI development, turned to a more human-centered technological breakthrough—personal computing. Nonetheless, Shakey is credited with taking the first slow, rattling steps toward the AIs and robots we see in use today, such as the Roomba vacuum.[18]

Shakey now enjoys a comfortable retirement on display at the Computer History Museum in Mountain View, California, and was inducted into the Carnegie Mellon Robot Hall of Fame in 2004 (along with *Star War*'s C3PO, among others), one year *after* UNIMATE and *Star War*'s R2D2 received the same honor.

Dad retired in 1980. By then, UNIMATE was off the job, too, in favor of newer robots. In 2014, TRW sold Thompson Products to Japanese manufacturer THK Rhythm. The parent company itself was bought by a German firm that manufactures sensors for self-driving cars. (Robot cars—Dad would have loved to get in on that action.)

Mom was right when she said that robots would take jobs away from humans. Dad was also right when he said robots would save workers' lives. A heated debate about whether automation destroys more jobs than it creates rages on. What we do know is that in manufacturing, robots now drive the pace of work, not humans.

We're now seeing the rise of "lights out" factories, so called because all the work can be done by machines in total darkness (and without heat and air-conditioning). But these factories are not 100 percent unmanned (or "unhumaned," to be more correct): a handful of workers are always on site to provide quality control.[19]

Why isn't every factory already a lights-out one? The answer lies in the type of situation that caused UNIMATE to occasionally malfunction: that is, having to handle exceptions to the normal routine of work. A HAL level of intelligence would be needed on the factory floor to deal with all the things that still go wrong. And even HAL couldn't handle the burden of conflicting priorities without going on an astronaut-killing spree.

As a result, we're still needed to pitch in and help when factory robots confront something unexpected. Possibly, not for much longer.

In 1979, AI and robotics paused, took a breath, and let Moore's Law do its magic, boosting the computing power it would need to meet its promises (including the household robots that Dad was so desperate to build). The stage was set for a new era when the balance of power would flip from AI to the human-centered philosophy of *intelligence augmentation* or IA. Computers—which, in the 1960s and even into the 1970s, were invisible to most of us—were about to become a dominant force in day-to-day work life that virtually wiped out entire industries and job descriptions within a few short years, introducing whole new ways of working, creating, and communicating.

As the age of the PC dawned, IBM worried that desktop computing might prove too daunting for anyone but confirmed techies (even though the PC was only marketed to businesses, not home users). Like the factories that hesitated to use UNIMATE until it was humanized by a guest spot on Johnny Carson, many businesses were fearful of computerization in the early 1980s.

An ad agency advised IBM to adopt an everyman persona that represented humor, simplicity, and humanity—everything so-called "Big Blue" IBM was not. The idea was to position the PC, not as a high tech innovation, but as a tool that would help businesses run so ultra-efficiently, employees would spend less time on office work and more on professional development, business building, and leisure. Unlike artificial intelligence and robotics, which were designed to replace human workers outright, the PC was sold as a way to make the workday more productive and enjoyable for people. It promised a utopian new world of paperless offices, shorter workweeks, and happier workers. And so, with hopes set high, began the era of IBM's Little Tramp.

The unfulfilled promise of artificial intelligence, which had unrealistically predicted thinking, mobile robots by the early 1980s, gave way to the era of desktop computing.

Chapter 3

A TRAMP IN THE AI WINTER

1985

> Are you ready to join your fellow countrymen (4 million Americans can't be wrong) and take home some bytes of free time, time to sit back after all the word processing and inventorying and dream the dear old dream? . . . There's a New World coming again, looming on the desktop. Oh, say, can you see it? Major credit cards accepted.
>
> Roger Rosenblatt, "A New World Dawns"
> *TIME* magazine, January 3, 1983

I was sitting on the other side of a desk from a guy whose job it was to lend money to small businesses. He could have been anywhere from fifty to a hundred years old: a sour, angry, frustrated, misogynistic, embittered man in a tweed jacket dusted with dandruff. He looked like a cross between a psychotic bookkeeper and the headmaster of a British boarding school whose curricula included brisk floggings after Scripture. I needed this guy's approval for a $5,000 bank loan, so I could buy a Zenith PC, my very first desktop computer.

I was twenty-eight, but looked younger. I had been self-employed for about a year as a hired gun for ad agencies that were weak in the writing department. Six months earlier, I'd left my job at Stone & Adler Direct, the Toronto office of a long-established Chicago direct marketing agency that counted IBM as its biggest client. Visitors to S&A

Toronto were greeted at reception by a life-sized cardboard cutout of Charlie Chaplin's silent movie character, the Little Tramp, wearing a bowler hat and leaning on a cane.

To humanize its brand, IBM had bought the rights to the character's image from the Chaplin estate—the real Charlie having died in 1977—and hired a mime to portray him in IBM's TV and print ads under the tagline "A Tool for Modern Times." The line was a nod to Chaplin's 1936 silent film, *Modern Times*, a slapstick comedy about the struggles of the Little Tramp to survive the dehumanizing effects of industrialization. It was an odd cultural reference for cutting-edge business technology, but IBM paid $36 million for the rights to the Little Tramp. Charlie was our boy.[1]

Although the agency had me writing about computers, day and night, I wasn't actually allowed to *touch* one: the keeper of the PC was the head secretary, Doreen, a shoulder-padded, big-haired Cheryl Tiegs wannabe who spent her lunches playing racquetball and strutted around the office the rest of the day wearing a lapel button that read LOSE WEIGHT NOW—ASK ME HOW, leaving in her wake a miasma of *Obsession*, a pheromone-laced perfume that conjured up visions of abandoned sex on sheepskin rugs.

My dimly lit, windowless office, across the corridor from the office slash living quarters of George the creative director, contained a desk, a chair, a wastepaper basket, and a filing cabinet. Not only was there no PC, but no typewriter.

On my first day on the job, I asked George: "What am I supposed to write on?"

On cue, Doreen arrived with a stack of yellow notepads and a box of ballpoint pens.

"Just write your copy and Doreen will type the decks on the word processor for you," George explained.

I told George and Doreen that there was no way that'd I'd be handwriting anything; even my rough drafts were composed at a keyboard, and my penmanship was so awful, Doreen wouldn't be able to read it.

Doreen stared at me as if I were a wad of gum she'd discovered stuck to the sole of one of her stiletto-heeled pumps. She whisked away the paper and pens and returned with an IBM brochure and a purchase order, sliding them across the fake wood laminate surface of my desk by the tips of her manicured hot pink fingernails.

"Just don't pick an expensive one," she suggested as she walked out my door, leaving me in a perfumed fog.

The next day, I arrived at work, avoided eye contact with the leering face of the cardboard Charlie in the lobby, and passed George's office, whose disheveled appearance suggested that he'd spent the night there. An IBM Selectric typewriter was squatting on my desk, next to a neat stack of white bond paper and a box of Wite-Out correcting fluid. I flicked the machine's on switch and felt a visceral pleasure at its steady electric growl.

As a sheet of bond sighed its way into the automatic paper carriage, my nose was tickled by a musky smell. Doreen was standing in the doorway of my office, arms crossed, face cemented in place.

"All good now?" she asked in a tone of voice that implied she really didn't give a shit, one way or the other, then clicked off down the hall.

That would blessedly be my longest interaction with Doreen until the day I resigned.

The irony of writing copy on a typewriter to sell personal computers wasn't lost on me, but seemed to be on everyone else. PCs, at least the ones marketed by IBM, were business machines. Setting down words on a page wasn't a computational activity but a lesser one called word processing, which only a professional secretary who had been trained how to touch-type was qualified to do.

I, however, having been told for years that my bad handwriting marked me out as a future juvenile delinquent, had taught myself to type in sixth grade. The IBM Selectric that sat before me was the

Cadillac of typewriters. You could even switch type fonts by removing a metal ball inside the machine.

Despite my crush on the Selectric, I was not immune to the allure of PCs. After all, I spend ten to twelve hours a day, six to seven days a week (depending on George's mood), writing about them. Fantasizing about the shorter work hours my ad copy was promising, I asked George: "Wouldn't it make sense for IBM to give us all a PC to work on?"

George laughed. "Do you know what one of those babies *cost*, Terri?"

Apparently keeping a twenty-eight-year-old woman copywriter chained to her desk day and night without overtime pay was more cost efficient.

Not that everyone believed the hype. In a 1982 letter to the editor of *PC: The Independent Guide to IBM Personal Computers*, Jack Rowbar, Traffic Manager for Plains, Georgia, griped: "What I'd like to know is who let this David Bunnell [*PC*'s editor in chief] out of his cage anyway. He's the same clown who once wrote that the Altair computer would 'control all the traffic lights in a major city.' I bought an Altair and all I could get it to do was change the lights on its front panel. Lord knows what claims he'll be making about the IBM."[2] And when *TIME* magazine departed from its yearly "Man of the Year" issue to declare the personal computer "Machine of the Year" in January 1983, Roger Rosenblatt wrote the lead editorial in the voice of a snake oil salesman:

> Ever see one of these before, mister? Yes, you. I'm talking to you, ma'am. Ever work one of these Commodores or Timex Sinclairs or Osborne Is or TRS-80 IIIs? How do you like them Apples? Just a joke, son. Good, clean fun. But you look so skeptical, like you're from Missouri, and I want to sell you one of these beauties, 'cause you need it and 'cause you want it, no matter what you say. Deep in your all-American heart (you are American, aren't you, pal?), you crave this little honey, which will count for you and store for you and talk for you, and one day it might even kiss for you (no offense, miss).

> Point is, it will save you time. Time time time. And we need all the time we can save.[3]

The promise of a George Jetson two-hour workday didn't pan out as advertised—at least, not right away. The learning curve was longer than the few seconds it took the Little Tramp to master his PC in the TV ads. One of my S&A colleagues spent long hours at the agency's sole PC, designing forms for estimates and timelines, work that had to be done at night when Doreen wasn't processing words. (More of a time-saver by far, he said, was the agency's first fax machine, which was easy to set up—just plug it in!—and started reducing turnaround times right out of the box.)

The IBM PC's screen mimicked the look and feel of their mainframe terminals: headache-inducing green text against a black background. It used nonintuitive commands, requiring that humans talk like computers rather than the other way around. Worse, you could only do one task at a time. When you wanted to switch from, say, creating a spreadsheet to writing a presentation, you had to stop and retrieve the right "solution" from your "library." (The word *multitasking* was coined with the introduction of Windows 95, which enabled PCs to run multiple programs simultaneously.) Software was stored on floppy disks, in boxes, inside cabinets. And despite the promise of going paperless, most people printed and filed hard copies of documents: if anything, word processing software like WordStar and WordPerfect *increased* the amount of paper piling up. And, to put it bluntly, a lot of that early software sucked.

In the 2011 issue of *Wired* marking the thirtieth anniversary of the launch of the modern PC, Christina Bonnington put the first IBM PC into perspective:

> It wasn't much by today's standards, or even yesterday's. The 5150 featured a 4.77 MHz 8-to-16 bit Intel 8088 processor. It was less powerful than other processors available from Intel and Motorola,

> but those were thought to be "too powerful" for a PC. IBM also gave the 5150 a full 64 kilobytes of RAM—expandable to whopping 256 kB—one or two floppy drives (your choice) and a monochromatic display. The 5150 was developed in less than a year by a team of 12 . . . and built using off-the-shelf components. Depending on how you configured your 5150, you'd shell out anywhere from $1,565 to $6,000 for one. That comes to $4,000 to $15,000 in today's dollars.[4]

PC novices worried about electrocution, crashes, and looking like a fool in front of their staff as they struggled to learn how to use the damn thing. In a 1982 article in *PC*, publisher Andrew Fluegelman claimed that he put in "200 or 300 hours" reading about software before he touched his PC:

> I think that what keeps 95 percent of people feeling weird about computers is learning how to boot the machine—how to get the thing running. You sit there in front of it, and you don't know how to get it started. You're afraid that it's going to snap at you or gobble you up or go up in smoke if you don't do the right thing. And I think the other great fear . . . is that when you get it running, it's going to lead you into some black hole that you won't be able to get out of. But when I started playing with my computer, I stopped feeling that it was a machine that was doing things to me; it almost instantly felt like an extension of myself. It was as though I had 2,000 extra brains grafted onto my skull. I really had that feeling—here are these extra brains, and they're really at my command, for me to string together or build together in any way that I choose.[5]

The same issue featured an in-depth interview with Bill Gates, who cautioned against overestimating the ease of use of personal computers: "We're still not at the stage where I'd tell my mother, or some naïve person, just to go out and buy one of these machines"[6]—an observation that shouldn't have pleased Old Ma Gates.

To gain acceptance among skeptical, scared business people (not to mention naïve ones), the PC needed a human, approachable face to soften the fear of all that new technology.

Enter the Little Tramp.

In TV spots, the Little Tramp used his IBM PC to sell iced cakes and fancy hats from storefronts in small-town America. No female Charlies appeared in those IBM ads, only bow-mouthed blonde secretaries whose hearts were won by Charlie's ability to create a spreadsheet in VisiCalc.

For every ad, direct mail letter, brochure, or take-one we created, the art director had to find the right ACI, or "Available Charlie Image." It might be Charlie making a spreadsheet of his skyrocketing sales, Charlie working on an ever-accelerating assembly line, Charlie doing inventory in his retail store, Charlie teetering on a pile of paperwork, or Charlie in his trademark pose, twirling his cane and enjoying all that extra leisure time provided by his PC.

Truth be told, the Little Tramp looked like a simpering dolt. But the subtle message in the ads was: If a mime in a bowler hat can do it, so can you!

IBM also relied on a tried-and-true marketing strategy known as FUD, standing for Fear, Uncertainty, and Doubt: the emotions you'd experience if you chose any other brand of computer. *No one ever gets fired for buying IBM* became a catchphrase.

In 1983, the same year as *TIME*'s "Machine of the Year" issue, IBM introduced a smaller version of the PC designed for home use. With an operating system incompatible with the IBM 5150 people used in the workplace, and a blank keyboard (you slipped a preprinted template over the keys depending on what task you wanted to do), PCJr had the look and feel of a toy. Maybe the engineers who created it couldn't imagine a world where you might like to bring home a bit of work from the 5150 in the office on a floppy disk and slip it into your PCJr to finish up a task after dinner. In 1983, civilized people did not mix office work with home work: instead, you "worked late at the office" (a euphemism for having an affair with your secretary).

PCJr flopped, but not before S&A made every attempt to move the crappy little suckers by setting up touch stations in Computerland stores.

The brief George handed me stated: *Invite customers to touch PCJr's keyboard.*

I raised my eyebrows at George who, once again, looked like he'd slept in his suit.

The word "touch" sounded so . . . *intimate*. Perhaps PCJr's screen should first invite the customer to come closer. To touch a key. (*Oh that feels good!*) Then another. And another. Pretty soon they'd be letting down their defenses. Getting comfortable. Experimenting with exotic punctuation. (If you could figure out where it was on PCJr's blank keys.)

Perhaps the customer would be stimulated to buy a PCJr. Check into a hotel. Order up a bottle of champagne. Light a candle or two.

Before you knew it, frenzied floppies would be sliding in and out of PCJr's disk drive, after which you'd light up a Winston and douse yourself with *Obsession* to cover the smell of silicon so your household appliances wouldn't know what you'd been up to. PCJr would have to pull itself together and catch a cab home, wherever that was, and hope you called the next morning.

I went into George's office to show him my hard copy. George laughed as he read my deck and scratched his beard. "Doesn't sound like IBM's brand voice but what the hell. I'll present it. But first write the *real* copy. I'll present that too."

In the end, George presented both versions to our client: the real copy written in IBM's business-casual brand voice and my sensual version (on a blank sheet of paper, no letterhead, no company name or logo). The client was amused (and for all I know, titillated). The sexy copy was quickly shredded. The real copy was programmed into the touch station.

And nothing happened.

Tens of thousands of PCJrs continued to languish at Computerland, unloved, untouched, and unsold. A year later, IBM finally pulled the plug on PCJr.

IBM had forgotten one of the long-held tenants of advertising. If nothing else works, sex sells.

Eventually, life at the agency wore me out. All that heavy perfume, the long hours, and a creative director who never went home from work and thought I shouldn't, either. I wrote a business plan (it said, "go to other ad agencies and see if they need a writer for hire"). I visited a creative director at Ogilvy & Mather who sold me a how-to book he had written about freelancing, and recommended I contact a guy he knew at another agency because "John appreciates a nice set of legs." Presumably, this was one of my selling points, along with a portfolio that already included a few award-winning projects.

My life as a hired gun had begun.

I handed a surprised and sleepy-looking George my resignation letter, picked up my last paycheck from stone-faced Doreen, leased my own Selectric typewriter from IBM, bought a cheap briefcase and answering machine, had business cards printed up, and never looked back.

Although I never actually used a personal computer at S&A, my IBM writing experience gave me a fake veneer of technical competence that proved useful: as more and more technology was introduced into the marketplace, I rarely needed to hunt for work. I worked on projects

***Ferris Bueller's Day Off* (1986)**

"I asked for a car. All I got was a computer," complains Ferris Bueller (Matthew Broderick) about the PC he uses to hack into his high school attendance records. Really? A teenager with an IBM PC XT in his bedroom, while I was struggling to get a bank loan for my lousy Zenith? The high price tag of an IBM PC XT made it an unlikely possession for a high school student. On the other hand, Ferris was a "righteous dude" whose bedroom was a playground of high-end electronic equipment. The XT was just one of the dream machines of teenage-hood in the 1980s.[7]

for laptops, cell phones, fax machines, photocopiers, electronic banking services, and automated voice mail—no answering machine needed! I wrote copy that pitched the Macintosh to scientists and engineers. Having learned the trick of creating a sense of comfort and trust between nervous customers and new technology, I was making enough to pay my half of the rent on a sprawling studio apartment over a store in a seedy stretch of bohemian Queen Street in Toronto that I shared with my artist boyfriend, Ron. I had some left over to put together the big-shouldered jacket with jeans creative-lady look that I needed to be taken seriously in meetings with chain-smoking, Scotch-swilling ad guys who still dominated the industry in those days, warhorses who would have been at home in season one of *Mad Men*.

By 1985, I was ready to take the next step: abandon my IBM Selectric for a PC with a pin-feed printer and WordPerfect word processing software. One or two of the graphic designers I partnered up with for freelance gigs had tossed away their straightedge rulers and paste-up boards for the graphical user interface of the Macintosh, a system even more expensive than the IBM, but much sexier.

My brother Rick, who had by then dropped the "y" in his name and become an engineer at an oil company, recommended I stick with an IBM PC, or one of the cheaper IBM clones; the Macintosh, exciting as it was, looked like a goner, having sold poorly after the initial excitement surrounding its launch in 1984, complete with an award-winning ad during the Super Bowl. One problem may have been that Jobs had fallen so deeply in love with his creation that the Mac had become more than a product to him. It was his work of art, and he wanted no one to tinker with it. Macintosh's graphical user interface operating system (Mac OS) was a closed one, making it harder for software designers to create programs for the Mac than for the IBM PC. As a result, fewer applications were written for the Mac at the time. There were other tech-based problems with the first Mac, but Apple primarily suffered from not having three letters in its name: IBM, resulting in Fear Uncertainty and Doubt. Better to stick with a brand that was tried, true, and used by over 80 percent of the business market. I went with an IBM clone.

It's hard to believe now that a basic desktop computer cost enough to require a bank loan. In 1985, $5000 would be the equivalent of a little over $11,000 today. And that was just for the poor girl's version of a PC I'd settled on: a clone of an IBM, manufactured by Zenith, one of the many electronics companies trying to get a taste of the desktop computing market.

Under the tagline "the quality goes in before the name goes on," Zenith was better known for manufacturing radios and TVs. Their PCs were less expensive than IBMs but not cheap. When you factored in the cost of a pin-feed printer and software, my cash flow required a loan, which I knew I could pay back in less than a year.

1984 Macintosh Ad

Orwell's dystopian novel, *1984*, was much on the minds of people in 1984. Jobs capitalized on the zeitgeist with an ad that portrayed Apple as a Valkyrie-like runner, smashing the face of a mind-controlling leader. The implication was clear: Macintosh was for creators and nonconformists. IBM was for drones. Written by Chiat/Day copywriter Steve Hayden and directed by Ridley Scott of *Blade Runner*, the ad was only broadcast once, during Super Bowl XVIII, and was named by *Advertising Age* as the best TV commercial of all time. In addition to the Macintosh, the "1984" TV spot launched the trend of producing epically creative (and expensive) commercials for the Super Bowl, turning the ads into as much of a draw as the football game itself.

Which brings me back to the bank in 1985, sweating out the application process under the rheumy gaze of that miserable loan manager.

He stared down at my application, back up at me, and shook his head in disbelief, depositing a fresh blizzard of dandruff onto his Harris Tweed.

"You got any collateral?" he asked, peering at me over his bifocals.

I swallowed hard. I wasn't quite sure what "collateral" meant.

"Such as?"

"Life insurance? Car? Investments?"

I shook my head no, to all three. I didn't want to admit that my most valuable possession was a ten-speed bike.

He glanced at my application again. “You have no credit history. No collateral. And you expect this financial institution to loan you $5,000.” He snorted and tapped the application. “Look, the only way I can approve you for this loan is for someone *with* collateral to cosign for you.”

My father accepted my request to drive the hundred miles to Toronto to vouch for me, without much comment. My mother was a different story. Although she’d been worried about me abandoning the security of my full-time job at S&A, she was proud to see her youngest daughter making it on her own, just like the character in *The Mary Tyler Moore Show*. When I repeated the conversation I’d had with the loan manager over the phone, she made a sound not unlike the Selectric powering up: a low, quiet growl.

Mom and Dad appeared in the uptown bank branch at the appointed time. Tweedy McDandruff was polite to them, even a little obsequious; he was probably surprised I’d managed to dredge up cosigners who didn’t look like they’d been pulled out of a back alley.

Dad was polite, but distant. Mom glared at Tweedy and said nothing, until after she’d signed her name under Dad’s on my loan application. She rose to her feet, looked down at Tweedy, and said: “Why the hell are you making life so difficult for our daughter when all she’s trying to do is start a business? I don’t like your attitude.”

He didn’t have a chance to answer because Mom went on to rip him a new one, in two languages. We left the office without handshakes or so much as *have a nice drive home*.

When I returned to the branch, about a year later, having paid off the loan in full, I learned that Tweedy McDandruff had been replaced by a cheerful young woman who wanted to know how she could help me. Too late. I had already moved all my accounts to a rival bank (where I went on to invest, set up a business credit line, open an education savings account for our sons, and apply for a mortgage).

I might have been better treated if the loan manager had been a robot. Although Tweedy McDandruff was probably just following the rules laid down for him by the bank in 1985, I'll bet he still had the discretionary power to make judgment calls. If you were persuasive enough, or went to the same college as the loan manager, or your brother's girlfriend's father was his high school football coach, or you flirtatiously implied that you might be willing to get together after the papers had been signed for a little post-fiduciary canoodling, I suspect he might have been inclined to approve the loan and waive pesky little details like collateral. If, on the other hand, you reminded him of his wayward daughter, or he was hungover, or he didn't believe that a woman in her twenties could run a business, or he noticed that her name ended in a vowel—a dead giveaway that her family might have immigrated from a country somewhere in southern Europe—or he just didn't like you, he might enjoy making the process unpleasant.

An AI, on the other hand, would have no such prejudices. An algorithm would know that (a) I had only been in business for a short time, but (b) I was incorporated, had a healthy balance sheet, a university degree, and (c) no debt—not even an outstanding student loan. The only borrowing I did was on my credit card, which I paid in full every month. And (d) I was a woman, which meant that, statistically, I was less likely to go bankrupt, women having a better track record than men in starting (and keeping) small businesses. Thus, the algorithm might shrewdly conclude, I might open my own agency, start an investment portfolio, purchase a car—hell, I might even take out a mortgage with their financial institution one day! I was a bundle of potential Return On Investment in a Crimplene dress and L'eggs panty hose. And I was asking for an amount that was chump change for a bank, even in 1985.

There's a pretty good chance the AI would have bumped up my interest rate a smidge and waived the collateral requirement, sending me off to Computerland with a $5,000 business line of credit to buy one its desktop cousins that very day.

Or, maybe not. The algorithm might have agreed with Tweedy McDandruff that I was a high-risk client and required a cosigner for the loan. Nonetheless, it would not have sneered at my application.

Knowing the value of customer loyalty, the AI might even have given me a complimentary toaster or calendar to show the bank's appreciation for my business.

Blade Runner (1982)

Released just after the end of the first AI Winter and at the dawn of widespread personal computing, Ridley Scott's noir adaptation of Philip K. Dick's novel *Do Androids Dream of Electric Sheep?* offers a glimpse at the past's vision of the future we've almost reached: 2019, when android-slaves known as Replicants sneak back to Earth from off-world colonies, where most humans now live, to find a way to extend their preprogrammed life spans. In addition to the super-strong (but childlike) Replicants played by Rutger Hauer and Daryl Hannah, who are so humanlike that they can only be detected through a psychological test known as a Voigt-Kampf, the film offers up PCs that look a lot like the ones on offer in 1981, as well as the usual futuristic flying cars, and Harrison Ford reading print newspapers in showers of acid rain that constantly drench a Los Angeles portrayed as more Asian than Latino. The past got the future wrong in so many ways, just as our futuristic films are no doubt doing today. Not many will have the staying power of *Blade Runner*, which has been called the greatest science fiction movie ever made. (Sorry, *Star Wars*!)

Given my youth, the AI would be aware that gaining my loyalty for the next fifty years could have been worth a hundred times the value of that loan, if not more.

Alas, no HAL-voiced bank bot came to my rescue to offer me a quick approval and/or a stress pill. That was still impossible in 1985. The AI Winter was coming.

Most people have never heard of the AI Winter. But mention it to an AI scientist or a roboticist today and they will sigh and give you a tragic look that may remind you of the agonized faces of Frodo and Sam journeying through Mordor.

The AI Winter blew in and out and in again, over the space of two decades, straddling roughly the late seventies to 1980, then blanketing the tech industry again from the mid-eighties to the mid-nineties. During these years, research in artificial intelligence virtually ground to a halt, which is one of the reasons we don't already have the robots predicted for the early twenty-first century by Isaac Asimov. If you want a short explanation for why the chill descended on robots, I'll answer with a much-quoted line from Karl Marx and *The Big Lebowski*: Follow the money.

From the earliest days of research into robotics and AI, private and public funding sought to capitalize on the next big breakthrough in automation. The biggest funder was the Defense Advance Research Project Agency (DARPA), the research branch of the US military formed after Sputnik caught America by surprise. Shakey, our mobile robot friend from the previous chapter received funding from DARPA because they saw potential for the robot to become advanced enough to work as a military sentry or astronaut. The simplest way to describe what happened circa 1975 was that after investing millions, if not billions, of dollars, DARPA and other investors lost faith in the potential for AI to be what we now call a moon shot—a highly ambitious, game-changing breakthrough. AI and robotics plunged into the widening chasm between promises and results: some prominent names wildly underestimated the time it would take to deliver a moving, thinking, seeing AI-driven robot, claiming in the late 1960s that it could be done within ten years. When robots like Shakey still didn't come close to displaying the problem-solving abilities, vision, and mobility of humans by the mid-seventies, confidence in AI began to erode. Even though progress was steadily being made, the funders had neither the patience nor vision to continue paying for research that might not lead to a fully useful robot or intelligent AI system for decades.

One of the biggest reasons for the gap between promise and delivery was a lack of computing power. Thanks to Moore's Law, AI researchers correctly assumed that they'd have more and more computing power at their disposal over time but, by the early 1980s, it still wasn't enough to create an AI as intelligent as HAL.

Another factor might have been bad timing. Whether, like me, you clearly remember the economic "circling the drain" sensation of the late 1970s or you've just heard about it in *That Seventies Show*, it was a time of belt tightening that gave rise to the word *stagflation*, meaning a stagnating economy coupled with rampant inflation. Jobs dried up. Profits fell. Personal earnings stalled. Interest rates skyrocketed. Worse, all this economic doom and gloom was set against a backdrop of an oil crisis and the political shock of the Iran Hostage Crisis. It's no coincidence that the 1970s was the decade of the disaster movie: like the giant atomic tomatoes of the 1950s, the 1970s pop culture thrummed with a constant, low-level paranoia.

One result of the tightening purse strings of that decade was the ramping-down of the space program. After you've landed on the moon a few times, gone on joyrides in a dune buggy, hit some golf balls, taken iconic photos of the blue marble of Earth against the blackness of space, ferried back tons of lunar rocks . . . what more should you do to entertain the folks back home?

With both funding and public enthusiasm for moon shots on the wane, NASA canceled Apollo missions 18, 19, and 20 to focus on launching the Skylab space station. From the get-go in 1973, Skylab was riddled with technical problems, eventually making a slow, much-anticipated fall to earth in July 1979. A space station doesn't fall out the sky every day, and the optics of the crash seemed ominous.

Like building the pyramids, space exploration requires a commitment to ultra-long-term goals. So do AI and robotics. Short-term thinking kicked in and personal computing seemed to hold more promise.

The first AI Winter began to thaw in 1980, due in part to President Reagan's "Star Wars" Strategic Defense Initiative. By 1986, forty start-ups in the United States were trying to commercialize AI. Computational linguist Jerry Kaplan, whose life had been transformed by watching *2001: A Space Odyssey* six times in the summer of 1968, wrote a program that would enable users to type questions to an AI in natural, human language. (He managed to complete this feat on an Apple II personal computer over his Christmas holiday in 1980.) This led to Kaplan cofounding Teknowledge, a start-up whose objective was to deliver AI systems that could automate the decision-making abilities of human business experts. Their process was to do intensive interviews with experts in various fields, build a database to house their knowledge, and write a problem- and logic-based program that would replicate the way the expert approached specific business challenges, a product known as an "expert system." Think of it as trying to replicate the brains of CEOs and accountants, *à la* the bottled heads in *Futurama*, and you'll have the basic idea. Teknowledge's CEO claimed that a business could save "as much as $100 million annually" by using an expert system instead of hiring high-priced talent.[8]

Teknowledge's investors were mostly venture capital firms with deep pockets. Expert systems ran on workstations that cost about $17,000 each, and a complete installation would run between $50,000 and $100,0000. But according to *New York Times* technology journalist John Markoff, Kaplan knew that the high-priced Teknowledge software would probably work just as well on a personal computer and actually wrote an expert knowledge program that would run on a PC—once again, over his Christmas holiday. (As Markoff drily pointed out, Kaplan "wasn't a big partier.") In the space of a couple of weeks, he'd created a product that could run an expert system—which had never fully lived up to the company's claims—for $80, instead of an average of $80,000. Although Teknowledge saw the value in the new product, they weren't happy that it undermined their business plan. Kaplan moved on.

Kaplan went on to join the Lotus Development Corporation,[9] where he helped develop Lotus Agenda, a database system designed to help PC users manage information.

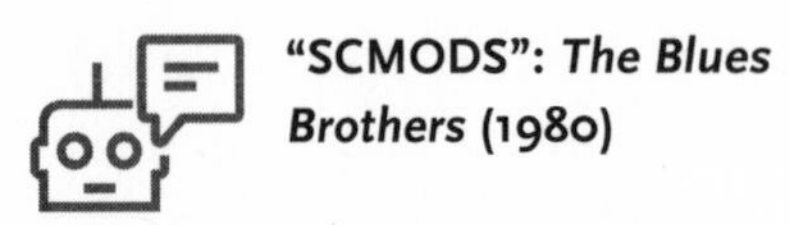

"SCMODS": *The Blues Brothers* (1980)

When the Bluesmobile is pulled over for running a red light, Elwood (Dan Ackroyd) mutters to Jake (John Belushi): "I bet those cops have got SCMODS: State County Municipal Offender Data System." Based on Mobile Data Terminals, the system wasn't a personal computer but an onboard dumb terminal connected to a police force's mainframe.[10] SCMODS identified Elwood as having a suspended license and fity-six moving violations, flashing the command "ARREST DRIVER—IMPOUND VEHICLE" on its screen, setting in motion one of the greatest car chases in movie history.

Business opportunity caused many AI scientists like Kaplan to jump ship on AI in to devote themselves to computer programming, a field sometimes called *Intelligence Augmentation* or IA. For some, this was also an ethical decision.

True believers in AI and IA have an animosity verging on contempt for one another that goes back to Norbert Wiener's *Cybernetics* at the dawn of the computer age. In the world of IA, you, the human, are in control; the computer is your tool, augmenting your knowledge and skills, but not replacing them—like the publisher in *PC* who said he felt the personal computer gave him "extra brains."

AI researchers, on the other hand, held that the true moon shot was to fully automate human intelligence and skills and embed them into systems that could replace humans, whether a delivery drone, a factory robot, or a self-driving car. (Having said this, today's AI developers emphasize that their goal is to assist humans and work alongside them, relieving them of dangerous or repetitive tasks. Despite ambitions to manufacture products in lights-out factories, move them to market on self-driving trucks, and store them in automated warehouses, the divergent goals of AI and IA appear to be merging, to some extent.).

There is a fairy-tale quality to the way the modern PC came to be. IBM was known in Silicon Valley as "Snow White," with its much smaller competitors called "the Seven Dwarfs." At least part of IBM's success may have been due to a businesslike, buttoned-down style that their customers found reassuring: IBMers had to conform to a strict, white-shirt and dark-suit dress code. They even had a company songbook.[11]

By 1955, there were about 250 computers in the world, each one worth about a million dollars. These behemoth machines required the floor space of a small house and frequently broke down, often because their inner recesses provided a cozy warm nesting ground for moths, a problem that gave rise to the term *computer bug*.

By the late 1950s, the valves inside computers had been replaced by smaller, more reliable transistors. Computers not only became faster and more robust, they shrank in size. You could put up a mainframe in your guest bedroom, if you felt like it.

Moore's Law predicted that computers would get smaller, faster, and more powerful over time. And lo, it came to pass. By the end of the 1960s, some computers had shrunk to the size of a refrigerator and were dubbed minicomputers (or microcomputers)—a name that evoked the mini- and micro-miniskirts then in fashion. DEC was "the IBM of minicomputers," with a billion dollars in annual sales by 1974. They continued to make smaller and smaller computers until they actually had developed something close to the size of a personal computer as early as 1973, but failed to realize the significance of what they'd done or to capitalize on it. This was a mistake that would later be repeated by another company, often spoken about by computer scientists in wistful tones reserved for Camelot or the lost land of Atlantis: the Xerox Palo Alto Research Centre, better known as Xerox PARC, a tech incubator full of wizards, geniuses, and seers who came up with, or advanced, some of the first, best, and stickiest concepts in personal computing, including[12] the graphical user interface (also called GUI or "gooey") of windows and icons, and a device called a mouse.

Originally demonstrated by its inventor, tech pioneer Doug Engelbart, at what came to be called "the mother of all demos," way back in 1968, the first mouse was carved from a block of wood. At a computer conference in San Francisco, Engelbart astonished the audience by using the mouse to jump from one part of a document to another, switch between documents, and expand and collapse lists of text—all at a time when the only way to give commands to a computer was on a teletype machine. The audience was slack jawed. But instead of rushing back to their labs to find ways to utilize the mouse, they said a collective "Oh wow cool man!" and forgot about it. An exception was Alan Kay, who had witnessed the 1968 demo, and went on to head up Xerox PARC.[13]

In the 1970s, Xerox positioned themselves among the Seven Dwarfs as a competitor to computing giant IBM. With their headquarters in Rochester, New York, Xerox's team decided it would be a good idea to situate their new research facility far away from the head office, where interfering managers might squelch the creative spirit of the young geeks and hackers they had hired for their brilliance.

You might ask why, then, did the best-known personal computer not bear the name Xerox? The simple answer is that they were not about developing commercial products, although they did come out with a storied PC—more like a laptop—called the Alto, and the computer language SmallTalk, both known as works of genius but never developed into commercial products.[14]

An alternate history has also been offered up: that the distance between PARC and its parent company in Rochester meant that the executives in charge of writing the checks were never involved in the process of approving these great ideas, nor is it likely they understood them, unless it directly affected their copier business.

What Xerox PARC did do was nurture some seeds into semi-maturity, then let them blow in the wind all over Silicon Valley, where they settled, took root, and grew in companies like Apple.

Steve Jobs visited Xerox PARC in 1979. In exchange for a million-dollar investment in Apple, he asked them to "open the kimono" so that he could gaze upon their technological pulchritude. According to Jobs' biographer, "Jobs bounced around and waved his arms excitedly . . . [He] kept saying he couldn't believe that Xerox had not commercialized the technology. 'You're sitting on a gold mine,' he shouted,"[15] sounding like a clever hero in a fairy-tale who talks his way into the inner sanctum of a carefully guarded castle full of secret treasures. The only thing missing was a unicorn. Jobs would eventually incorporate as much as he could of Xerox PARC's unmonetized magic into the Macintosh.

To understand how computers went from the size of refrigerators to phone books, we have to take a quick detour to Intel, where in 1969, Marcian "Ted" Hoff was asked to develop a chip for a Japanese-manufactured pocket calculator, a product that dominated high-tech manufacturing for a few short years. Hoff wouldn't accept that his microchip would be limited to helping people do arithmetic calculations and spelling out the word HELLO in digital letters, one of the many fun ways to turn a calculator into a plaything. He insisted on developing a microchip for computers, not calculators, nonplussing both his American employer and Japanese client. Fortunately, both were smart enough to realize that Hoff was onto a good thing. Thanks to Hoff's insistence on over-delivering, the Intel 4004 computer chip was introduced in 1971. Now computers could be made so small they'd actually fit on your desktop!

Recycled 1950s military computers as set design

The IBM AN/FSQ-7, a massive, decommissioned military computer, was a frequent stand-in for generic computers in a raft of movies and TV shows starting in the late 1960s. Originally built to track and intercept enemy bombers, the AN/FSQ-7 showed up in *Hawaii Five-o* (1969) *Sleeper* (1973), *Columbo* (1975), *Battlestar Galactica* (1979), and many, many others, including Dr. Evil's submarine lair in *Austin Powers Goldmember* (2002). While these "guest appearances" gave people a notion of what a computer was *supposed* to look like—room-size metal boxes covered with gauges and blinking lights—the AN/FSQ-7 was a dinosaur compared to the DEC minicomputers of the late 1960s and early personal computers, like the Altair and Apple 1, which were emerging in the early 1970s.[16]

The first personal computer to use the chip, the Altair 8800, was developed, not by Snow White or by the Seven Dwarfs, but by a tiny company called MITS (Micro Instrumentation Telemetry Systems), in Albuquerque, New Mexico. The owner, Ed Roberts, a former officer in the air force, started MITS to sell radio transmitters for model airplanes by mail order, later moving up to manufacturing calculators. He worked out of a strip mall in a defunct restaurant formerly called The Enchanted Sandwich Shop, the sign still hanging on the front of the building.[17]

Roberts took his electronics know-how to the next level by developing the Altair, which took its name from a planet visited by the *Enterprise* on an episode of *Star Trek*. He pitched the Altair, not to businesses, but at home users known as hobbyists. In 1971, hobbyists didn't mean a bunch of guys messing around with gluing together vintage airplanes or setting up elaborate model railroads, but were what we would now call "hackers." They were into computers for the sheer love of it. The tiny, low-cost Altair offered them the same amount of computing power as the 1950s behemoths and it was priced under $700. The Altair didn't have any applications—yet. But it didn't need to. The hackers would look after that.[18]

A young hacker in Seattle, Bill Gates, had been employed to find bugs in DEC minicomputers while he was still a sixteen-year-old high school student. According to one tech history book, Gates "could, by typing just fourteen characters on a terminal bring an entire [DEC] TOPS-10 operating system to its knees."[19] Gates and his partner Paul Allen wrote a BASIC program that could be used for the Altair, the first sale for a company the two of them had formed that they called Micro-Soft.

One particular group of hobbyists jumped on the opportunity presented by the cheap, small, and powerful Altairs. Starting in 1975, they had begun to get together to swap information and ideas about microcomputers, build cool stuff, and show off for one another. They called themselves the Homebrew Computer Club, a loosely allied gang of self-described subversives who were inspired by the San Francisco counterculture of the 1960s and believed in the democratization of computing—or, as they put it, "computers for the people." They weren't into computing for profit, but for fun.

According to club member, Steve Wozniak, "The theme of the club was 'Give to help others.' The Apple I and II were designed strictly on a hobby, for-fun basis, not to be a product for a company. They were meant to bring down to the club and put on the table during the random access period and demonstrate: Look at this, it uses very few chips. It's got a video screen. You can type stuff on it. Personal computer keyboards and video screens were not well established then. There was a lot of showing off to other members of the club. Schematics of the Apple I were passed around freely, and I'd even go over to people's houses and help them build their own."[20]

Most of the Homebrew Club's members—which grew from thirty to over one hundred—were tech types who worked in Silicon Valley's many companies; Wozniak was an engineer with Hewlett Packard, while another hobbyist, Wozniak's high school buddy Steve Jobs, worked for the video arcade company, Atari, maker of games like Pong and Space Invaders. Jobs would often sneak Wozniak into the Atari factory at night to play games without having to feed the machines coins. The games started to suggest to the two of the machines what a personal computer might be like.

Wozniak wrote that, for games, "You need sound: when the ball hits the bricks, ping: when you lose, *ehhhhh*. So I put a speaker in . . . These turned out to be common features for the personal computers that have come out since that time. We weren't quite the first to offer a keyboard and video output, but we were close . . . We made the first built-in cassette port so you could use a cheap cassette recorder to load your programs and store them. We had started to set standards for what have come to be known as low-cost personal computers."[21]

For Wozniak and Jobs, what had started in 1975 as a way to show off to the others in the computer club quickly became serious business. Jobs wanted to start a company as successful as Atari and nagged Wozniak to give up his day job at Hewlett Packard. At one point early on in Apple's history, an ad agency tried to convince them to change their name to something more high tech. They resisted. Given the amount of advertising dollars IBM would eventually invest into humanizing themselves with the Little Tramp, Apple's decision to stick with what Wozniak called a "friendly, healthy, and personal" name was a prescient one.

In less than ten years from the founding of the Homebrew Computer Club, Jobs introduced the world to the first Apple Macintosh, amazing his audience with computer graphics that spelled out the words "insanely good computers" in cursive writing. The Mac's mechanized voice (which now sounds shockingly like Steven Hawkings' synthesized voice) went on to introduce Jobs as "someone who has been like a father to me."

Oddly, although the Apple was as much a personal computer as the IBM 5150, the term "PC" quickly became attached at first to the IBM brand, and then to any desktop computer with an MS DOS operating system. Apple desktops would take on the name "Macs" with the introduction of the first Macintosh in 1984, eventually giving rise to the "Mac versus PC" meme.

The IBM Personal Computer was introduced in 1981 after a rapid-fire, secretive program to build a personal computer in record time with a team of twelve and the services of Bill Gates' company, Microsoft (they'd dropped the hyphen), to write the operating system known as PC DOS (MS DOS for non-IBM systems). According to Gates, development of a new IBM product usually took a minimum of four years but with the Apple II and a raft of other personal computers from the Seven Dwarfs already on the market, IBM knew it had to act fast to stay competitive. According to a seminal history of the birth of the personal computer, *Fire in the Valley: The Birth and Death of the Personal Computer*, the resulting machine was:

> . . . almost conventional from the standpoint of the industry at the time . . . inventor [of the Sol and Osborne personal computers] Lee Felsenstein got his hands on one of the first delivered IBM PCs and opened it up at a Homebrew Computer Club Meeting. "I was surprised to find chips in there that I recognized," he said. "There

weren't any chips that I *didn't* recognize . . . IBM is off in a world of its own. But, in this case, they were building with parts that mortals could get."[22]

As uninspired as the first IBM Personal Computer might have been, it revolutionized the marketplace simply *because* it was an IBM, a brand that stood for rock-solid reliability. Its presence built a new product category among nontech-savvy consumers and sparked opportunities for software developers who wrote a raft of programs to run on the new system.[23] As if to remind personal computer buyers that the Apple II was there first, Apple ran an ad in 1981 headlined "Welcome, IBM. Seriously."[24]

Two other technologies that appeared in and around the AI Winters are worth mention. One of them is the humble ATM, almost forgotten today for being one of the first computerized everyday devices most people had ever touched.

While we don't think of ATMs as high-tech breakthroughs anymore, they're on my personal list of "greatest inventions, 1956 to 1986." If you think this an overstatement, I invite you to jump in a time machine and join me at a bank branch, circa 1976, in an interminable line as you attempt to cash your paycheck. You might be in that line through your whole lunch hour. Come back after work, you say? Good luck. Bankers' hours in the 1970s were roughly 10 a.m. to 3 p.m. It was a time of scribbled deposit slips, customer frustration, and cranky tellers who insisted on putting a hold on checks, just because they could. Personal service could be really, truly awful and massively inconvenient.

The first ATM, the Bankograph, was introduced in 1960. Even though it spit out a printed photograph of the deposit as a receipt, people didn't trust it. In the words of the Bankograph's inventor Luther Simjian: "The only people using the machines were prostitutes and gamblers who didn't want to deal with tellers face-to-face, and there were not enough of them to make the machines a worthwhile investment."[25]

ATMs didn't catch on until a literal perfect storm occurred in 1977: Citibank invested $100 million in installing ATMs all over New York City, and a massive blizzard dumped seventeen inches of snow, shutting down the city (and banks) for days. In the teeth of the storm, New Yorkers braved the streets to withdraw cash from ATMs, an inadvertent public relations coup for Citibank because it highlighted the fact that you could now bank anytime without having to deal with bankers' hours or cranky tellers. The bank's ad agency capitalized on the situation by running ads headlined "Citi Never Sleeps," with photographs of New Yorkers trudging through the storm.[26]

Another new technology that first appeared in the 1980s and is now evolving into a major player in AI is the mobile phone. The technology for handheld phones had been around for decades, with a prototype developed by Motorola in 1973. Mobiles as we now know them—called cellular phones, because of the cell technology used to transmit signals—appeared in the United States as car phones in 1983.

Cellular phones would end up being a big part of my copywriting life, starting with a direct mail package selling Canada's first cell phone network, a plum assignment given to me by the creative director at Ogilvy & Mather who'd remarked on my legs. Despite being a sexist jerk, he turned out to be a good client.

Just as with PCs, I'd be writing about cell phones years before I actually owned one, and decades before the introduction of the iPhone and the cascade of smartphones that followed. In the words of Chris Atkeson, Professor of Robotics at Carnegie Mellon University, who we'll meet in the chapter, "The Good Robot": "The smartphone is a robot's head. Now we're just trying to give it arms and legs."

Before we return to the world of robots and AI, let's jump back once more in time to me in 1986. Bank loan in hand, I bought my Zenith PC and it turned out be a complete lemon, with frequent crashes that had me returning to the store again and again for fixes (and eventually an outright replacement).

Within a year, I had a Macintosh Plus—necessary, because I was working directly with graphic designers who had all jumped from manual design tools to the Mac. Since Macs and PCs were, at that point, incompatible with one another, I needed both systems, depending on who I was working for.

The switch to the Macintosh came at a turning point in my life: I married my boyfriend Ron, had a baby, and moved to a gold-mining town above the Arctic Watershed, almost 500 miles north of Toronto, while still freelancing for major clients in Toronto, Montreal, and San Francisco in an era before mere mortals like me had access to the Internet or email. To move copy to clients, I used a modem, and developed a relationship (strictly business) with the only Ricoh fax machine repair guy in northeastern Ontario, a vast wilderness area about the size of Texas and Montana combined. The local Apple dealer in the town where we lived called itself Blade Runner, because the owner had a side business selling hunting knives. We moved back to Toronto in 1994, just in time for the introduction of the World Wide Web, email, and Windows 95.

By then, personal computers had become a commodity, with IBM just one of many Windows-based PCs in the marketplace, and customers comfortable enough with the technology that they no longer needed the reassurance of the IBM brand. In the space of a few short years, PCs had entirely changed the workplace and the wider world, wiping out jobs for typesetters, typographers, couriers, and secretaries like Doreen, to name but a few of those displaced by desktop computers. The new world promised by the *TIME* article had finally arrived.

IBM left the field to its competitors to focus once again on Enterprise-level systems, selling its ThinkPad line of laptops to the Chinese computer manufacturer, Lenovo, in 2005. Although they stopped being the household name in PCs for personal use, I often find myself in a thicket of the matte black ThinkPads beloved by my corporate clients, while I sit like *Sex and the City*'s Carrie Bradshaw with my elegant silver MacBook Pro, both sides of the boardroom table still observing the creative-verus-number-cruncher brand loyalties of our tribes after all these years. IBM's artificial intelligence Watson is now the invisible, games-playing brain that could be even more of a game changer than the first IBM PC, moving us closer to the new dream of truly intelligent machines that can think and learn.

The Little Tramp no longer walks among us, and "a tool for modern times" was replaced long ago with the single word THINK, a throwback to IBM's motto from the 1950s and 1960s.

A vintage IBM THINK sign sits on my desk beside me, inspiring me as I process words and graphics on my Mac, whose founder's tagline over thirty years ago was *Think Different.*

Electronica, Synth, and TechnoPop

Computerization made itself heard in the robotic sound of techno-infused pop music throughout the 1980s and 1990s. The German band Kraftwerk recorded *Computer Land* in 1981 with tracks entitled "Pocket Calculator," "Numbers," "Home Computer," and "It's More Fun To Compute." Dressed in yellow jumpsuits, the industrial punk band Devo appeared on *Saturday Night Live* in 1978 to cover the Rolling Stones' "Satisfaction" in a stilted, mechanical style that sounded like the vocals were being performed by a computer—or a robot. And the biggest hit on Thomas Dolby's 1982 album *The Golden Age of Wireless* was the retro-punk pop song "She Blinded Me With Science."

Chapter 4

FUN AND GAMES WITH THINKING MACHINES

2018

I, for one, welcome our new computer overlords.

Scribbled by Ken Jennings as the answer in Final Jeopardy in the last round of the *Jeopardy!* IBM Challenge

I'm in my third time-wasting hour on Facebook, playing my favorite word game app. My usual opponent, Sandra, is not taking her turn, although with a smartphone, tablet, and laptop at her disposal, I don't understand why she ever needs to be offline. While she's away, I feed my addiction by playing against the robot thoughtfully provided by the app for solo players. I set the difficulty level to "10," select the US dictionary, and we're off.

I start the game with PEEVE for fourteen points. Robot responds with WEDGY for thirty. I fight back with a fifteen-point BERM. Robot resists with an unexpectedly weak WOWS for seventeen. Pouncing on the Triple Word Score that Robot has unwisely opened up for me, I spell SIX for thirty points. Robot retaliates with a forty-four-point QUARTO. Not to be outdone, I parasitically riff off the "U" and "A" in QUARTO to spell ME, UM, and AE for twenty-two points.

Robot fights back with what algorithms do best, dredging up an obscure word that only the geekiest human would know: THECAE for twenty-eight points. That's when my game starts to slide. Two moves later and Robot delivers the final blow with PENNATE for twenty-eight points. I'm sunk. Robot passes its final turn and still winds up beating me by a humiliating seventy-two points. I can almost see the smirk on its virtual face.

Robot has instant access to every word in English (and several other languages) in all standard dictionaries. All I have is a human brain, sharpened by higher education and crossword puzzles, but the worse for wear from too many Friday nights sipping Pinot Noir while Springsteen blasts from my top-of-the-line-in-the-1970s Mesa stereo speakers. Like the twenty years, worth of outdated iMacs and MacBooks in my basement, I can barely boot up my brain's obsolete operating system.

But Robot does not always win: I sometimes beat it by as much as 100 points. How is this possible? Robot consistently uses words that give it the highest possible scores on every turn, ignoring the fact that it sometimes sets me up for points-rich triple-word scores. Robot has no strategy. No intuition. No *soul*.

However, my occasional triumph raises the question: am I playing Robot, or is Robot playing me, by losing just often enough that I'm tricked into feeling as if I'm playing a human being? Is it programmed to sometimes behave as if it's sleep deprived, stressed, or distracted by its ailing West Highland Terrier, the way my friend Sandra is?

Robot isn't just playing with me, it's learning how quickly I make words under pressure, whether my scores drop late in the day when my brain is getting tired, and the number of times in a row it can beat me before I give up in frustration. It can then apply what it learns about me to other humans. That's important, because as disembodied algorithms, AIs don't have any context for what it means to be human. They didn't grow up playing Twister or Rummoli, and don't have memories of throwing temper tantrums when their big sisters beat them at Monopoly ten times in a row. If AI-enabled robots are going to work and play with us, they need to understand the quirks, foibles, gaps, and even the strengths of human intelligence. (And yes, we still do have some.) Right now, the only place that AIs can get to know us is where many of us increasingly spend our time: online. Game apps, social media, and YouTube videos have all become part of the AI's Dummy Guide's to Understanding Humans.

Similar to the way children learn by playing, games-playing AIs will help give robots the ability to navigate their way in our maddeningly unpredictable world. There's no better place to see this in action than in Pittsburgh, a city of suspension bridges, deep-fried pickles, self-driving Ubers, and world-class robotics research. There's even a robot repair shop at the airport.

I'd come here with my husband Ron as my second pair of ears and eyes to visit the Carnegie Mellon Robot Hall of Fame and watch an AI named Libratus learn to sort out truth from deception by throwing down against four professional poker players. En route, we caught a glimpse a Ford Fusion beetling around a corner near the University of Pittsburgh, its rooftop LIDAR rotating like the bubble on Gort and Klaatu's flying saucer. Frustratingly, we never got to ride in one. Every Uber we summoned was manned—or more technically, womaned. (Actually the self-driving Uber is manned, too; state laws require a "safety driver" at the wheel who can take over in case the car needs a carbon-based life-form to pitch in and help.) I asked one Uber driver what she thought of the self-driving cars. She snorted: "You'd never catch me ridin' in one of those things. I picked up the head of Uber when he came to Pittsburgh and I said, 'I guess you're trying to put me out of business,' and he said, 'Not for another ten years.'"

"One of the roboticists at Carnegie Mellon told us that they're here because autonomous cars have a hard time with bridges," I said.

She snorted again. "This city is nothing but bridges."

This is the one of the reasons Uber set up its Advanced Technology Center in Pittsburgh, where 446 bridges (including one bright yellow suspension bridge named for favorite son Andy Warhol) hang gracefully over three major rivers and a vast network of valleys and ravines, a constant flow of traffic moving over them. For autonomous vehicles, Pittsburgh must look like a giant Pac-Man game, but instead of outmaneuvering cherries and ghosts, they have to learn to navigate Pittsburgh's unpredictable traffic, construction sites, and weather.

Not only autonomous cars, but AI-enabled robots of all kinds will have to learn to make snap decisions in an ever-changing, illogical world. You and I take calculated risks all the time. It might be whether to cross a busy intersection as a traffic light is changing, give the cold shoulder to a chatty stranger, or trust a salesperson who claims we look fabulous in red. Whatever: we use a lot of brainpower making minute-to-minute decisions based on clues we may not even be aware of, a form of intelligence known as "executive function." Somewhere deep inside our brains we're aware that cars run red lights, strangers can be dangerous, and sales people sometimes lie; we can't take what we see or hear at face value. To have any hope of functioning in the real––as opposed to virtual––world, artificial intelligence needs to pick up on the same clues we do, without enough data to be 100 percent certain of their choices. Another word for that type of decision making is gambling. And what better way for an AI to learn to beat the odds than by playing 120,000 hands of Heads Up No-Limit Texas Hold'em against four professionals, with real money on the line?

It's the opening day of the Brains Vs. Artificial Intelligence Match at Rivers Casino, right next door to the Carnegie Science Museum, home to Roboworld and the Robot Hall of Fame, a must-see for a robotophile like me: Gort is here, along with Robby from *Forbidden Planet*; Will Robinson's robot, B-9, from *Lost in Space*; and R2D2 and C3PO. HAL 9000 looks disappointingly like the air-conditioning unit in a cheap motel—although, when you think of it, how *do* you portray an artificial intelligence? Humans are accustomed to interacting with AIs as the disembodied voice of a Siri, Alexa, or some other digital assistant. A better representation might have been a loop of actor Douglas Rain voicing HAL's best lines or slurring the words to "Daisy Bell."

While the life-sized replicas of pop culture robots stand in their niches, the real ones, including Dad's friend UNIMATE, are only shown in reproductions on the Wall of Fame. The rest of Roboworld is filled with working robots (well, *mostly* working; a few are out of order). There are robots that fill prescriptions, paint abstract art, react empathetically to your facial expressions, and sternly request you to *please move* if you block its path. My favorite is Hoops, a bright orange industrial robot with a forklift-style arm, an obvious descendent of UNIMATE, right down to the nets surrounding it on all sides to protect onlookers from being bumped, scooped up, and tossed through a hoop.

In a past life, Hoops worked in a factory welding car bodies, but he (it?) has been reprogrammed to sink baskets with 98 percent accuracy. Lacking the sensors of more advanced robots, Hoops feels around with his pincer-like arm to locate basketballs on the floor or inside a trough. When Hoops finds a ball, he scoops it up and tosses it over his head—or where I'd imagine a head would be, if Hoops had one—into a basketball net, then spins and dips his arm in a series of movements that looks suspiciously like a victory dance. Part of the point behind Hoops (beyond entertainment value) is to show how good manufacturing robots have become at doing repetitive tasks. After a few minutes of watching Hoops put the ball in the net and give that little bob and weave of joy, it's impossible not to find the big orange lug weirdly endearing. Hoops has been shooting baskets at Roboworld and in the museum's traveling road show since 1996. He's a pretty simple robot by 2018 standards, programmed to play one game with a clear win/lose outcome, over and over again. Which, when you think about it, is what the high-priced talent in the NBA do all the time. It's also a good way to think about how robots learn. Hoops has been programmed to do his thing, the way you'd program a computer—as much as Hoops' movement seem like some guy on a basketball court, the robot is not thinking. More advanced robots can actually learn by imitation, something that seems a little freaky. It's one of those post-AI Winter truths that convinces some of us that robots are turning into big, unruly, silicon-based teenagers, who will eventually rebel against their human parents, steal the keys to the car, and turn us into batteries, like in *The Matrix*. No worries about that from a single-minded old robot like Hoops.

I search for a Hoops T-shirt in the science museum gift shop, but no luck. (Good museum, disappointing merch.) With my cash still in my pocket, we stroll along one of the paths following the banks of the Ohio River, with its three Uber-defying bridges, to the aptly named Rivers Casino. It's a short walk into a very different world from family- friendly Roboworld.

The casino has a miasma of indoor smoking and the otherworldliness of a neon cathedral full of furtive supplicants in tracksuits, hunched at arcade-like slot machines branded with everything from Kate Winslett in *Titanic* to characters from *Harry Potter* and *My Little Pony*. Even the long bar running through the middle of the casino provides a game console at every barstool. Signs are plastered everywhere that read: THINK YOU HAVE A GAMBLING PROBLEM? CALL 1-800-GAMBLING.

It's hard to believe that if you're in a casino in the middle of a Wednesday afternoon in early January, that you *don't* have a gambling problem. The hollow-eyed players hunched in the shadows of one-armed bandits do not look like they're here for fun. The question of whether they have a problem seems to answer itself.

"Maybe we should play a couple of games," I suggest to Ron, thinking of the unspent cash, earmarked for robot souvenirs, in my purse, and the time I won (and lost) a stack of cash at roulette in a Salzburg casino on our honeymoon. My human brain remembers winning big in the past––why not try my luck again?

Ron, on the other hand, uses his brain's executive function to calculate the risk to our bank account. He gently steers me away from the slots: "Let's just find the poker room, shall we?"

We arrive to find that the Brains vs. Artificial Intelligence Match has just gotten underway between four human champions and a baby (of sorts)—Libratus, an artificial intelligence that, like Hoops, was built for one purpose only: to win at Heads-Up No-Limit Texas Hold'em. Libratus has never taken on a human before, let alone four of them. Its poker play is not based on trying to mimic the experiences of expert humans (à la Teknowledge's expert systems of the 1980s) but by using its own algorithms to analyze how its human opponents are playing the game. It even knows how to bluff. One other thing Libratus has in common with the human players: it refreshes itself overnight, learning to improve its play by analyzing its daytime games while its opponents are offline (also known as "asleep").

What does a Texas Hold'em-playing AI look like? Does it watch its opponents' eyes for a "tell"? Can it cheat? We expect to see something like PokerStars on ESPN with steely-eyed players in baseball caps sitting in rich Corinthian leather club chairs at green felt tables. Or the high-stakes game in the movie *Lock Stock and Two Smoking Barrels,* with a gang of British thugs making half-million-dollar bets, smoking cigars, and drinking Scotch at a table set up in the middle of a boxing ring to a retro-cool sound track by The Castaways.

The Brains vs. AI Match is not quite that exciting; Libratus and its four human opponents are playing on PCs with their cards projected on screens hanging on the wall behind them, as a handful of spectators look on. Not far away, the casino's regulars play poker the old-fashioned way, oblivious to the history-making human-machine tournament unfolding in their midst. The entire match will last twenty days.

In the first Brain vs. AI match in 2015, a Carnegie Mellon University AI named Claudico lost to the human players, although the game actually ended in a statistical tie. Still, Claudico failed to win bragging rights, so this time, it's personal: the CMU computer scientists are determined to be the first to build a supercomputer that can defeat four professionals at the "no limit" version of Texas Hold'em. There's real tension in the air among geeks and gamblers alike. That, and a lot of cigarette smoke drifting in from the other side of the casino.

Libratus' DNA can be traced back to the chess-playing IBM supercomputer, Deep Blue, which faced off against grand master Garry Kasparov in Philadelphia in 1996 and again a year later in New York. Ever since that famous tournament, games have been the benchmark for measuring how well artificial intelligence performs against human intelligence. Today it seems intuitively obvious that a supercomputer could beat a human at chess; twenty years ago, not so much.

The stakes were high for IBM. Despite having done pioneering work in artificial intelligence, the company abandoned AI research in 1959, possibly over concerns about damaging the company's reputation should it become linked with technologies that destroyed jobs.[1] But when it became clear in the late 1980s that business opportunities would grow along with the power of the still-new Internet, IBM wanted in. Deep Blue was built to prove that AI could deliver on the promise of human-like machine intelligence, and that IBM could deliver AI. Now they just needed a human opponent smart enough to compete against a supercomputer.

***WarGames* (1983)**

Matthew Broderick plays a high school student who uses a phone modem to hack into a computer to change his grades—shades of the character he'd play five years later in *Ferris Bueller's Day Off*—and ends up in a real-life nuclear death match with a military supercomputer with near-disastrous results. If a computer asks, "Shall we play a game?"—don't answer by suggesting "Global Thermonuclear War." Just don't.

If anyone fit the bill, it was Garry Kasparov, still regarded not only as the greatest chess grand master of his own time, but of *all* time. He accepted IBM's challenge to play Deep Blue good-humoredly, regarding the match as a friendly competition to help advance computer science research. In other words, he didn't take Deep Blue seriously as an opponent.

The Philadelphia match went to Kasparov, as he confidently predicted it would. In an article he wrote for *TIME*, he explained how Deep Blue's computational power had been offset by its lack of what might be called intuition. After Deep Blue played "flawlessly" to win the first game, Kasparov used the computer's own dogged inflexibility to work against it:

> . . . my overall thrust in the last five games was to avoid giving the computer any concrete goal to calculate toward; if it can't find a way

> to win material, attack the king or fulfill one of its other programmed priorities, the computer drifts planlessly and gets into trouble. In the end, that may have been my biggest advantage: I could figure out its priorities and adjust my play. It couldn't do the same to me. So although I think I did see some signs of intelligence, it's a weird kind, an inefficient, inflexible kind that makes me think I have a few years left.[2]

In fact, Kasparov had *less* than a year before he came face-to-face with a souped-up version of Deep Blue, officially named "Deeper Blue." With help from an American grand master who had once played Kasparov to a draw, IBM programmed the computer to function more like a human chess player that could adapt its strategy to its opponent's moves.[3] Kasparov agreed to a rematch in May 1997 in New York, waving off IBM's offer to split the $700,000 prize between winner and loser 60/40; he insisted on winner take all.[4] The illusion that the rematch was a "friendly competition" vanished. Instead, it was hyped like a boxing match. Long gone was the friendly, human face of IBM's Little Tramp. New York City was plastered with posters of a close-up of Kasparov's penetrating stare over the headline: "How Do You Make a Computer Blink?"

I remember the Kasparov-Deep Blue match as a major media event, part of the wallpaper of tech news that dominated the years leading up to the turn of the century. Y2K anxiety was rampant. Echoing the nuclear fears of the 1950s, predictions were made that in three short years the so-called "Year 2000 problem" would mean the end of the world as we knew it. The problem was an idiotically simple one: since the dawn of the computing age, years were programmed as two numbers rather than four (e.g., "99" instead of "1999"), which meant that the year 2000 would register as "00"—the year 1900. To computers, it would seem as if time had jumped backwards by a century, sending their logic-driven processors into a HAL-style meltdown. Every computer, from desktop PCs, to bank and government systems, to networks that control power, water, and military defense, would crash simultaneously at midnight. Electrical grids would fail, cities would go dark, email and phones would stop working, and planes would fall from the sky. Overnight we would find ourselves crawling helplessly through the wreckage of a digital Armageddon. These doomsday predictions spawned not only groups of survivalists hoarding water, food, and gold, but a massive amount of work for programmers racing against the clock to fix the problem. As the clock ticked down to 2000, a lot of techies got rich.

I, too, profited from Y2K anxiety: one of my best-ever assignments was writing comic books for a large corporate client that wanted a simple way to get its branch operations to pay attention to its head office's requirements for standardized, Y2K-compliant computer platforms. Instead of yet another sternly worded memo, the comic books featured the adventures of an entertaining superhero named Standards Guy who could fly in and solve your department's Y2K-related issues, using the same boring procedures everyone had been ignoring up to that point. (The comics worked gangbusters and my client weathered Y2K just fine, thank you very much.)

In the midst of all this gloomy tech news, the Kasparov-Deep Blue competition floated like a life buoy for humanity. Y2K showed that computers weren't so smart after all—they couldn't even handle a date change without human intervention! It was far from a done deal that Deep Blue would triumph over the brilliant Kasparov.

To recapture the zeitgeist of the late 1990s, I turned to Google and YouTube—time-machine tools undreamed of twenty years ago—to look at media coverage of the match. My favorite photo shows thirty-four-year-old Kasparov in a snappy three-piece suit, head in hand, gazing at the board; behind him, rows of leather-bound books and duck decoys are lined up on wooden shelves, as if the match is going on in an exclusive men's athletic club. In fact, they were playing on the thirty-fifth floor of the Equitable Insurance Center in midtown Manhattan, with a few hundred spectators in the first-floor auditorium watching the play on a video feed and many more following on the Internet; IBM's site would crash under the strain of so much online traffic.[5]

Opposite Kasparov, an IBM programmer stares at a laptop, which acted as his communication link to the two Deep Blue supercomputers six floors below. Between the men are a timer and two small flags: the stars and stripes on Deep Blue's side, an Armenian tricolor for Kasparov who, although a Russian national, was of Jewish Armenian decent. Deep Blue's ancestry went back to Carnegie Mellon University, where in 1985 two graduate students built a chess-playing computer called ChipTest, later renamed Deep Thought for the computer in *The Hitchhiker's Guide to the Galaxy*. Both computer scientists were hired by IBM in 1989 to work with the team that developed Deep Blue over seven years.[6]

Deep Blue was actually not one supercomputer but two, each about the size and shape of a big refrigerator. IBM researchers acted as the computer's eyes and hands, typing Kasparov's moves into the laptop and moving pieces on the board as Deep Blue instructed. Kasparov never laid eyes on Deep Blue. "I have no information about my opponent," he complained.

Kasparov easily won the first game of the match. But in the second round, Deep Blue opened by sacrificing its pawn, a move that startled Kasparov; it seemed too "sophisticated" and "human" for a computer, unlike anything he had seen Deep Blue do before. With Deep Blue playing its side of the game in private, Kasparov grew suspicious that he was being duped in some way. Since exploiting an opponent's quirks, weaknesses, and superstitions were all part of high-level chess, Kasparov's team believed that IBM was playing a psychological game.[7] The computer even made a move that reminded Kasparov of his former arch-nemesis, the grand master Anatoly Karpov. At one point, Kasparov became convinced that Deep Blue had been switched with another computer between games.[8]

What Kasparov was actually up against was *Brute Force Computing*: Deep Blue's 256 processors examined 200 million possible moves *every second.* Kasparov, on the other hand, had to depend on his past experience to look ahead at what moves might follow. He couldn't hope to foresee the number of moves that Deep Blue was able to consider.

But winning and losing a chess tournament isn't just about coldly calculating moves: it also involves psyching out and wearing down your opponent, something that Kasparov could not do to a machine. Unlike Kasparov, Deep Blue was incapable of becoming fatigued or frustrated.

Over the course of the match, Kasparov changed his mind about the supercomputer being switched. Instead, he had come to believe that Deep Blue's game was being played by a human, possibly a grand master, aided by the supercomputer. In the press conference after the second game, Kasparov used the term "hand of God" to explain his loss: chess-speak for "cheating."

Games three, four, and five were all draws, although Kasparov almost quit midway through; he was simply too tired to go on. In the sixth and final game, Kasparov lost badly, bitterly explaining that he played "an emotional game against a machine with no emotions."[9]

The match would wrap up with Kasparov's team accusing IBM of sabotage and strong-arm tactics. At the final press conference, Kasparov sounded frustrated, while IBM's team came across as smug. No one came out looking good except IBM itself: according to a documentary about the match, the company's stock rose by 15 percent in one day when Deep Blue won.[10] Despite Kasparov's request for a second rematch, the supercomputer never played again. One half of Deep Blue is now in the Smithsonian.

It seems that Kasparov's team was right: the whole point *had* been to crush him psychologically. But why would IBM put so much time and money into developing a supercomputer whose sole purpose was *to beat one man in a chess game?*

Kasparov was no ordinary man, but a brilliantly gifted human. Beating a lesser opponent might not have been accepted as proof that Deep Blue truly could play with (and defeat) the best brain humanity could pit against it, a notion that still dominates research today, in which AIs are always pitted against "the best of the best."

IBM may have seen the Kasparov-Deep Blue match as a Turing test that could prove that a supercomputer could make decisions the way a human would: using common sense, based on a series of rules that had been programmed into it. In the AI world, this approach is called "Good Old-Fashioned Artificial Intelligence" or GOFAI. It may work in a controlled situation like a chess match, with its highly regulated, logical play, but less so in the hurly-burly of the real world. For robots of 2018, something beyond the limits of GOFAI was needed—a new type of artificial intelligence that would allow AIs not just to follow rules laid down by their programming, but to learn, on their own, from their own experiences and humans themselves.

As for Kasparov, who seemed to be the sacrificed pawn in humanity's decades-long competition of brains versus AI, he was far from crushed. Instead, the grand master's first defeat opened a new chapter in his life, inspiring him to come up with ways for humans and computers to work together cooperatively. (More on that later.)

IBM continued to pit faster and more powerful chess-playing computers against grand masters, but none captured the public's imagination the way Kasparov and Deep Blue did. Like the moon landing, the excitement was sucked out of the accomplishment once it had been achieved. Sure, a computer could play a game of logic and patterns like chess. But could it intuit

"Kasparov vs. Machine" Superbowl XXXV Pepsi commercial (2001)

Four years after the loss to Deep Blue, Kasparov parodied himself in a commercial, in which he outplays a supercomputer at chess and scornfully announces that "all machines are just wires, nuts, and bolts—they're stupid by nature." After the game, every machine he comes across (a security camera, elevator, floor polisher, and soda machine) conspires against him. The spot ends with Kasparov flying backwards through the elevator doors, which close on him ominously, while the soda machine displays the words YOUR MOVE with a smiley face on its digital screen.

like a human? Understand language without taking every statement literally? And why couldn't a computer play a game directly with its human opponents, instead of having a programmer act as a middle-man? That type of skill required something that hadn't been developed yet—a computer that makes decisions based on more than facts and logic, and speaks naturally, the way that HAL communicated with the astronauts on Jupiter 1 or the computer on the *Enterprise* responded when Captain Kirk barked "Computer!"

When the clock ticked over at midnight on New Year's Eve, 1999, and the lights stayed on, water flowed from taps, and planes didn't crash, the world took a deep breath, and moved deeper into a world driven by the Internet. The first decade of the 2000s saw the growing power of search engines and the launch of Facebook (2004) and YouTube (2005), quickly followed by Twitter (2006) and other social networks, turning Google into a verb and raising expectations for what computers could do. IBM was ready for a new Grand Challenge that would capture the world's attention. Executive Charles Lickell discovered it in a restaurant in Fish-tail, New York, one evening in 2005. When a crowd of diners suddenly left their tables to hurry into an adjoining bar, Lickell remembers: "I turned to my team and asked 'what's going on?' It was very odd. I hadn't really been following *Jeopardy*, but it turned out it was when Ken Jennings was having his long winning streak, and everyone wanted to find out if he would win again that night, and they'd gone to the bar to see."[11]

As he watched the game, it dawned on him that the skills needed to win at *Jeopardy*—tackling answer-and-question problems, not in the manner of a search engine, but the way a human did, and in natural language—was exactly the kind of challenge IBM should be taking on. He pitched the idea to the people at IBM Research, who were skeptical: this went way beyond chess, to a level of AI no one had ever attempted before. Not to mention, putting a supercomputer on a TV quiz show seemed a little gimmicky.

In case you've somehow managed to miss it since it went on air in 1964, *Jeopardy* is a quiz show with a twist: answers are given to the contestants, who try to beat their opponents by buzzing in first with the correct questions. The category titles can sound mysterious: for example, when I checked the the *Jeopardy* game categories for March 7, 2017, they included "Becky and the Good Hair," "Pack Lunch," "Elemental Words," and "A Watery Border." Some questions are straightforward and factual, running the gamut of history, geography, pop culture, science, politics, and what-have-you. Others are weirdly convoluted, forcing contestants to make sense of cryptic play-on-words and puns. If the contestant who buzzes in first gets the question wrong, the other two contestants get a shot at it. And they do all this in less than three seconds. At the end of each game, contestants bet as much (or as little) of their winnings as they like, knowing nothing more than the category of the final answer. When the final answer is revealed, they have thirty seconds to come up with the correct question while some goofy theme music plays. *Jeopardy* isn't just a game of general knowledge, but of calculating the odds of having the right answer. Speed is a big factor, too: you can be wrong a lot of the time and still win the game by being the first to buzz in with the right question often enough.

Contestants give off a whiff of bookishness and can be more than a little . . . shall we say . . . geeky. When describing IBM's AI, *Jeopardy* superstar Ken Jennings wrote self-deprecatingly that it "has lots in common with a top-ranked human *Jeopardy* player: It's very smart, very fast, speaks in an uneven monotone, and has never known the touch of a woman."[12]

Developing an AI that could compete against the equivalent of a "quiz show Kasparov" meant giving it abilities that went far beyond Deep Blue's brute force computing, or even a search engine like Google; winning at *Jeopardy* wasn't just a matter of culling through millions of facts and picking the right one, but finding an answer that made sense out of the puns and plays on words that are part of the game. Somehow, the AI would have to mimic our ability to figure out double entendres, like Groucho Marx's quip, "One morning I shot an elephant in my pajamas. How he got into my pajamas, I'll never know." And on top of all this, the AI had to be able to speak like a human.

A research team spent four years, held scores of practice rounds, and underwent two auditions before they got their AI on *Jeopardy*. They named it (or, to use the pronoun preferred by Alex Trebek, "him") Watson, after Thomas J. Watson—IBM's president from 1914 to 1956.[13]

Atari Breakout (1975) vs. DeepMind (2015)

Steve Wozniak and Steve Jobs created the game Breakout for Atari as a version of Pong: instead of batting a ball back and forth between two players, a single player bounced a cursor against a brick wall until it finally broke through. Thirty years later, Google's DeepMind AI taught *itself* to play Breakout. Nothing was preprogrammed into the system; the only information given to DeepMind was its scores, so that it could judge for itself how well it was doing. At first, the AI was totally inept at the game, missing most of the shots, but overnight, it learned to play Breakout perfectly. Since then, DeepMind has proven to be a fast learner at almost fifty other classic Atari games.[14] Known as *machine learning*, this method of teaching AIs how to learn on their own sounds a lot like a never-ending, rainy weekend in a beach cottage with a gang of bright kids keeping themselves entertained, 24/7, with no parents around to tell them to get the hell to bed.

The logical, commonsense approach of Good Old-Fashioned Artificial Intelligence wasn't enough to turn a computer into a *Jeopardy* champion. Watson had to be much more than a fact finder; it had to *learn* to play through trial and error and by watching other games, similar to the way a human contestant would. Using 2800 processors the size of ten refrigerators, Watson was fed ten million documents, downloaded from encyclopedias, Wikipedia, *The New York Times* archives, and the entire IMDB movie database (to name but a few information sources). Because Watson was not connected to the Internet while playing *Jeopardy*, it could only access what was in its own database: it couldn't "Google" an answer. IBM also provided Watson with thousands of examples of *Jeopardy* answers and questions, training the AI to hunt for patterns from past games to refine its own answers.

Despite all this machine learning, Watson couldn't be right 100 percent of the time. For example, there would always be puns and other illogical turns of phrase that would stump it. So Watson learned to calculate the odds. If Watson was sure of its response, it bet more; if less certain, it bet less.[15]

Who's reading your mail?

Machine learning of the type used to develop Watson was also adopted by the US Postal Service to train an AI to sort mail by reading addresses on envelopes. They did this in a similar way to Watson learning how to come up with winning answers to Daily Doubles: the postal AI was provided with a vast number of examples of each letter of the alphabet, in every type font, as well as many different handwriting samples. Eventually, the AI caught on to the patterns of forming letters: an "A" was an object with a pointy thing up top, legs like a tent and (sometimes, but not always) a crossbar between them.[16] Eventually, it learned to recognize the shapes of every letter of the alphabet.

Although Watson could speak in a soft, faintly hollow male voice similar to HAL's, it couldn't see or hear: the *Jeopardy* answers and other contestants' questions were transmitted to Watson by text message. Like Deep Blue, the supercomputers that ran Watson were kept separated from its human opponents because of the roar of the cooling system of its ten racks of Power 750 servers.[17] So, to give Watson a physical presence, IBM designed an avatar of a blue ball crisscrossed by forty-two "thought" lines (forty-two being a reference to *Hitchhiker's Guide to the Galaxy*'s supercomputer Deep Thought, which after 7.5 million years of computing, came up with that number as the answer the ultimate question to "life, the universe, and everything"—but unfortunately, no one remembered the question.)

In February 2011, the producers of *Jeopardy* pitted Watson against their two greatest contestants: Brad Rutter, who had won $3 million on the show, the highest winnings in the game's history, and Ken Jennings, whose seventy-four-game winning streak was still unbroken.

The *Jeopardy!* IBM challenge took place on a set custom-built for the match, with the show's longtime host Alex Trebek at the helm. Brad Rutter said, it was "a bit of a John Henry thing for me. I kind of like the idea of going up against a machine and showing that humanity has something else going on that maybe you can't, at least not yet, reproduce digitally."[18] Jennings wrote after the fact that: "Unlike us, Watson cannot be intimidated. It never gets cocky or discouraged. It plays its game coldly, implacably, always offering a perfectly timed buzz when it's confident about an answer. *Jeopardy* devotees know that buzzer skill is crucial—games between humans are more often won by the fastest thumb than the fastest brain. This advantage is only magnified when one of the 'thumbs' is an electromagnetic solenoid triggered by a micro-second precise jolt of current."[19] After playing three hard-fought games over three days, Rutter came in third place with $21,600, Jennings in second place with $24,000, and Watson way out front with $77,147. Jennings summarized the game in a way that was weirdly reminiscent of Kasparov: "Game over for humanity."[20]

But Watson wasn't perfect: in game two, it incorrectly provided the question "What is Toronto?" in response to the Final Jeopardy answer, "Its largest airport is named for a World War II hero, its second largest for a World War II battle." (The smaller airport in my hometown is named for a World War I flying ace, Billy Bishop, the larger one for a Nobel Prize winning prime minister, so it's hard to understand how Watson could have blown that one.) Both Rutter and Jennings correctly answered, "What is Chicago?" But because Watson wasn't confident in its response, it bet very little.

Watson also continued to struggle with riddles and puns: under the category "Computer Keys," the answer was "Proverbially, it's where the heart is." Watson was stumped; Rutter got it right: "What is home?"—as in the home key on a computer and the proverb "Home is where the heart is." But Watson's problem understanding puns wasn't big enough to prevent the AI from winning, because it bet conservatively whenever it wasn't sure of the meaning of a *Jeopardy* answer.

Unlike Deep Blue, the victory wasn't the end of Watson, but the beginning. The AI had commercial applications written all over it. Watson's combination of problem-solving and natural language skills has since turned the AI into a tool for medical diagnosis and marketing data analytics (i.e., new ways to sell you stuff).[21] And that, of course, is only the beginning of its resume. IBM is now training Watson for jobs in education, financial services, and the Internet of Things.

Both Garry Kasparov and Ken Jennings commented on the oddity of playing against a computer, not just because of its brainpower, but because of the physical advantages of being a machine that doesn't fatigue or get discouraged or, as Kasparov put it, get "psyched out."[22]

Playing an AI is not simply like playing a very smart human: it is like playing a genius who doesn't get hungry, thirsty, tired, nervous about how it looks on camera, or in urgent need of a bathroom. Over the years, I've watched this benchmark of 24/7 performance applied to us human weaklings in the workplace. We are expected to be always on and always connected, never unable to go to work even in our so-called off hours. This may be a reality that we can blame not only on email and smartphones, but the tireless performance of our machines.

With chess and *Jeopardy* conquered, the next challenge for AI was the ancient game of Go. As senior tech writer Cade Metz wrote in *WIRED*: "Go is epically complex. An average turn in chess offers about thirty-five possible moves. A Go turn offers 250. After each one of those moves, there are another 250. And so on. This means that even the largest supercomputer can't look ahead to the results of every possible move . . . In order to crack the game; you need an AI that can do more than calculate. It needs to somehow mimic human insight, even human intuition. You need something that can learn."[23]

Before Google introduced AlphaGo, if you'd asked a tech-savvy Go player how long it would take to develop an AI that would be able to beat a human grand champion at the game, their answer would probably have been ten years. It took two.

In January 2016, AlphaGo beat three grand masters in the DeepMind Challenge Match. Google embargoed the results for a couple of weeks, probably just to give us humans time to take a chill pill and let sink in what they'd just achieved. Why such a big deal? Unlike Deep Blue, AlphaGo couldn't use brute force computing to look at every possible move because the number of possible board combinations in an average 150-move game of Go is greater than the number of atoms in the universe.[24] Like Watson, AlphaGo had to bring something humanlike to the game. It had to *learn from its mistakes*. According to project manager David Silver, AlphaGo's neural network would look at a model of what a human player was likely to do in a specific situation, but it was also trained to look at "less probable moves . . . searching more deeply and analyze things in an introspective way."[25] Google says their goal for DeepMind is to become a "generalized AI system that could build on its knowledge and apply its learning to anything."[26] Translation: AlphaGo could learn to make the kind of decisions that snag humans a corner office and a seven-figure salary. Doctors, lawyers, CEOs . . . will AlphaGo replace them all one day, or like Watson, become the ultimate executive assistant? Only Google knows for sure.

Outside of tech circles, most of us heard the news of AI's latest games-playing triumph without surprise: *of course, the AI always wins! It's just another sign that humanity is doomed*. In the years since Deep Blue,we seem have given up on our ability to dominate a world where machines can outthink us. We pessimistically expect to be replaced or exploited by AIs. But will this actually happen?

As I would later learn from a specialist in *deep learning*—the newest way that AIs learn to learn—AIs are amazing but they have weak spots (mostly because they learn everything from us). And, as Kasparov would go on to demonstrate, the real opportunity in the future might be in buddying up with AIs, rather than assuming they'll replace us outright.

You might think that no game is off limits for an AI. But in 2011, a tech analyst confidently stated that a game of chance and bluffing—poker—would *definitely* be outside of the realm of possibility for an AI like Watson. Why is poker so much harder than chess or *Jeopardy*? A poker-playing AI would need to take advantage of human frailty; it would have to know its opponent's game, and make moves based on what it thought likely they were likely to do, which wouldn't always necessarily be logical. And the more players who took part in the game, the less well AIs would perform.[28]

Which brings us back to the Brains vs. AI Match in Pittsburgh. After twenty days and 120,000 hands of Straight Up No Limit Texas Hold'em, Libratus beat out *all four* of the human professional players—the first time such a feat had been achieved. To ensure that the players were doing their best, the humans were playing for money (as opposed to Libratus, which was playing for glory).

Like Watson and DeepMind, Libratus can learn to learn, and spot patterns that let it predict outcomes with a super level of accuracy, a skill that could be applied to medical diagnoses, cyber security, and negotiating business deals.[27] But, for the moment, Libratus focuses on honing its skills against other poker-playing computers. *AIs playing against one another,* rather than against humans, has become the new normal.

Arnold the AI takes on DOOM

The first-person-shooter video game, DOOM, challenges players to "kill or be killed" while fighting 3-D demons from hell. Players have to chase and escape from enemies, read maps, and cheat their way out of close calls. In 2016, Carnegie Mellon computer science graduate students Devendra Chaplot and Guillaume Lample used learning techniques similar to Deep Mind's to train their AI "Arnold" to outplay human opponents. Arnold eventually went on to play DOOM against other AIs from Facebook and Intel. Wily, quick-thinking AIs like Arnold might someday help self-driving cars adapt to sudden changes in weather, traffic, and road surfaces.[27]

Games help AIs learn to deal with the uncertainties of the human game of life. But with AI-driven robots in development that can recognize faces, call humans by name, and give personalized care—and keep each of us distinct from animals, inanimate objects, and other humans—I started to wonder: how do robots learn about us as individuals? And do they have any of the frailties humans do? For answers, I went to a hip coffee shop in Toronto's Kensington Market area to meet computer scientist and deep learning specialist Xavier Snelgrove. Still in his twenties, Xavier is Chief Technology Officer and Cofounder of Whirlscape, creators of Dango, an AI-driven emoji product designed to make digital communication more expressive. Think of the number of times you've added a smiley face or a sad face to an email to provide the emotional context you can't convey through body language or tone of voice, and you'll understand the appeal of Dango. Xavier explains that their main target market is teenagers, but I think of how many times I've found myself wasting minutes browsing through potential emojis as a final *mot juste* for an email or text message.

Xavier's girlfriend Cheryl is at the coffee bar too, working on her laptop. When I ask her what she does for a living, she tells me that she's a designer. "What kind?" I ask, thinking she'll answer graphic, fashion, or industrial.

"I design health-care experiences," she says, a reminder that the digital native generation is different from mine. Even our job descriptions exist on either side of a chasm.

Xavier explains that the AI industry constantly uses games to set benchmarks and mark breakthroughs because it's easy to tell when they're succeeding. "There always has to be signal for success or failure," he explains. "Winning and losing at games gives an AI that signal."

To answer my question about how machines learn to recognize individual humans and their environments, Xavier swipes the touch screen of his phone to show me rooms designed by an artificial intelligence. All of the images are a bit out of focus, as though they were snapped by a drunken party guest reeling through the hallways of the designated hotel for a destination wedding. They have the faintly anonymous, blandly designed look of messy hotel rooms: rumpled beds, floor-to-ceiling curtains, cluttered surfaces. A few suggest something more intimate, possibly even romantic: Victorian-era beauty tables with crystal bottles and lace trims. At least that's what I read into them: everything is a bit indistinct. These could be glimpses of rooms from one of the dreams I often have of houses I've never visited, my subconscious creating an interior from childhood memories or magazines I might have flipped through at my hair stylist's. That's probably a good metaphor for these rooms—dream rooms, imaginings. None of them exist outside of the collective minds of artificial intelligence: they're like children flipping through a stack of old home décor magazines, then drawing picture after picture of their own designs, shouting "*Room! Room! Room!*" And where did they derive these data-driven ideas of "room"? From you, me, and everyone else who posts photos of rooms (or our lives inside them) on Facebook, Instagram, and anywhere else that the AI hive mind can crawl over.

"They're still a bit small and blurry, but getting better all the time," says Xavier, sounding a bit like a proud papa trying to downplay his gifted offspring's talents.

As Xavier puts it, these interiors were designed by "non-embodied algorithms"—in other words, a HAL-like AI that interacts with us through our devices—phones, cars, our home-heating system, smart appliances, and who knows what else. Maybe one day mobile robots like R2D2 will come equipped with this type of insight too. At first, they will see a room and just think "room." Eventually, they'll see a room and know it's "Terri's room," simply by the stacks of discarded copy paper and empty bottles of rosé.

"What's the difference between machine learning, deep learning, and Good Old-Fashioned Artificial Intelligence?" I ask.

Xavier grins: if there's one thing he enjoys talking about, it's AI learning systems.

"GOFAI is a rules-based system that uses a commonsense approach to problem solving. What would *we* do? The assumption is," he says meaningfully, "that we know ourselves. That we can give the AI a set of rules based on common sense. But machine learning means writing down many examples and letting the machines come up with the rules. Deep learning is based in neuroscience. It uses the human brain as a metaphor and a source of inspiration but does not directly try to replicate a brain. We don't tell the AI principles: we show it examples and let it come with its own principles."

Xavier goes on to explain that he uses supervised learning to train AIs: "It gives you random answers and you tell it when it's wrong. It's a remarkably simple learning algorithm. Our own nets only take about a week to train."

Sitting in this laid-back, funky coffee bar, with Cheryl designing experiences beside us, and Xavier talking about "training" disembodied algorithms, I feel like I'm at the edge of something large and unpredictable, possibly unruly, definitively disruptive, and quietly seductive. AIs are starting to sound . . . er . . . *alive.* These overachieving, hyper-intelligent virtual children may still be in kindergarten but they sure learn fast.

"So playing a game—getting answers right and wrong—is how an AI learns?" I ask. "Are there ever times when they *don't* learn from their mistakes?

Xavier nods. "Sometimes the AI will start to assume that because it had a particular outcome once in the past it'll have that outcome again. That's called over-fitting. 'Oh I remember, when I did X, I lost' so . . . we have to use huge data sets with lots of examples, so big that the AI couldn't possibly just memorize them. That forces it to learn principles."

"Over-fitting sounds like what people do. You might incorrectly learn something from one experience that shouldn't apply to all experiences," I point out, thinking of the roulette win on my honeymoon that made me confident I could hit the jackpot again at Pittsburgh's casino.

Xavier nods. "Right. Over-fitting is a challenge for self-driving cars when they confront problems they've never seen before."

I remember the self-driving Ubers challenged by the Andy Warhol Bridge. It turns out that teaching an AI how not to be fooled is important. How do you prevent someone from sabotaging cars by putting up fake signs to misdirect them? How can you make sure that the self-driving car recognizes that someone is trying to send it in the wrong direction?

"In other words, self-driving cars need a set of principles to get around in an unpredictable environment rather just relying on past experience?"

"That's right," says Xavier. "The research in this field is moving so quickly you just don't know where it'll be in five years. Something brand new is generative techniques. If you ask an AI to draw a truck, you'll get an abstract idea of a truck based on all the pictures of trucks on social media, and the web, but now we can also use generative *adversarial* networks. For example, if I wanted to generate a picture of a human face, I can train two neural networks, but I'll train one of them to be *really good* at telling when it's fooled by the other one."

I stare at Xavier. I'm trying not to show how completely freaked out I am by the idea of AIs trying to put one over on one another. Now they sound less like kindergarteners, and more like high-stakes gamblers trying to bluff one another. I can almost smell the whiskey and cigarette smoke.

"You're training one neural network to try to fool another one."

He nods happily. "Yes, it's an adversarial process. They go back and forth trying to develop something realistic. It's like they're inventing a new game and playing it together."

AIs are trying to create a world that mimics the competitiveness and one-upmanship of our human world. While they don't—at this point–think like humans, they are attempting to understand how we think, trick one another, and occasionally cooperate. They are even helping one another recognize individual people and objects so they can to distinguish between (for example) a small child, a large dog, and a washing machine (and treat each of them appropriately).

Xavier explains that a technique called "transfer learning" allows an AI to transfer data from one part of its (let's call it) brain to answer a question it has never confronted before: "For example, if you showed an AI a picture of a zebra, and it had never seen one before, it would be able to transfer the idea of, say, a 'stripey cow horse' from other data sets and eventually figure out what it is."

"So, do our interactions with AI change the way that we act and think as humans?" I want to know. "Have we given up on ourselves too soon?"

"Yes," said Xavier firmly. "We now think of both chess and Go as mechanical games, instead of poetic ones. We defer to AIs. We trust them to be smarter than us. But there are always biases in the data sets we choose to show AIs. We have to remember that. For example, the University of Shanghai trained a vision net to look at mug shots in order to be able to spot a criminal type just by looking at a face. There was a swift response from the larger AI community that this was wrong because the mug shots were not an objective truth. It's like when you see photos being 'enhanced' in shows like CSI. All the AI is doing is filling in the sections of the photos that can't be seen from photographs it gets from its data sets."

I'm gobsmacked. The enhanced photo of so many TV shows—and a critical scene in the original version of *Blade Runner*—are not really enhancements of tiny details in photograph, but an AI making an educated guess at what's in the photo. A digital hunch, let's call it.

Before I leave, I tell Xavier about my ongoing word game battles against Robot.

"Sometimes I beat Robot by one hundred points," I tell him. "Other times I lose by that much. Do you

The Internet is made of cats (yes, really)

In 2012, Google X's computer scientists built a neural network of sixteen thousand computer processors with one billion connections, and in an "unleash the hounds" moment, set it loose to browse randomly selected YouTube video thumbnails. The goal for the AI was to learn how to identify faces and objects. One of the things it proved to be really, really good at identifying was the faces of cats.[30] Why? Because so many of us (including me) post pictures of our adorable fuzzy-cutie-pumpkin-toes on YouTube, Facebook, and other social media. Those kitty pictures are helping drive the deep learning algorithms that AIs used to learn to recognize objects. Thus, as I've always suspected, cats rule the Internet and dogs drool.

With that in mind, I've made a point of posting many, many photographs of my cat Echo on social media. I want to subtly influence the algorithms to prefer petite, all-white cats with yellow eyes, dainty paws, and missing ear tips (the result of frostbite from a kittenhood living rough on cold Canadian streets, until she was rescued). I like to think of Echo as my little contribution to deep learning.

think it's letting itself lose, sometimes, just so I don't get disheartened and stop playing?"

Xavier doesn't let me down easy. "It has access to every available dictionary so I don't think there's any doubt about that. It's definitely letting you win."

My friend Sandra is offline. *Again.* And Robot is really beginning to piss me off. While I sit for minutes at a time scanning my letters, Robot forms its words in a microsecond. Annoyingly, the game refuses to accept *niqab, munchie, rez, da,* or *zen* as legitimate words, while it happily allows Robot to suck up points with *qi, za, ut, qat,* and *ka.* It's all I can do not to tell Robot, "fuck you and the algorithm you rode in on."

AlphaGo, Watson, and Libratus show that AIs are starting to learn in the same way a human does—by experience. And when Robot gets better at that, it'll know better than to open up triple-word scores to me early in the game when the Zs and Qs are still up for grabs.

LEGO *Mindstorms* (1998)

The humble LEGO brick is so iconic that most of us think of it as too classic to tinker with. But with the rise of video and computer games in the 1990s, LEGO went through an existential crisis. How could its build-your-own town and fire engine sets appeal to the new generation of digital natives? The result was a "build your own programmable robot" kit, inspired by AI pioneer Seymour Papert, whose book *Mindstorms* gave the toy its name. Although the product targeted LEGO's typical customers—twelve-year-old kids—it turned out that 70 percent of the people playing with *Mindstorms* were adult hobbyists of the Homebrew Computer Club–type who had discovered that the set was much more than a child's toy. Along with motors, gears, axles, wheels, and about seven hundred bricks, the $150 set included a software application that could be used to program a microcontroller-based brick, the RCX (Robotic Command Controller), which had the same computing power that put a man on the moon. When a Stanford University graduate student reverse engineered the RCX brick and posted the code on the Internet for all to see, LEGO decided to let the hackers spread the word and added a "right to hack" to the *Mindstorms* software license. Soon, *Mindstorms* discussion groups started popping up and software developers were writing applications for the toy. Robots known as Mindstorms MOCs (My Own Creations) flooded the Internet, sparking LEGO League Tournaments for *Mindstorms* geeks. Two of the MOCs were an assembly plant that custom built a LEGO car and a working vending machine that dispensed soda and change.[31] LEGO *Mindstorms* was inducted into the Carnegie Mellon Robot Hall of Fame in 2008.

For a moment, I experience a shiver of queasiness known as The Uncanny Valley, that creepy feeling that a robot is beginning to act like a human.

At that moment in 1997 when Garry Kasparov stood up from the chessboard and walked off in disgust after losing game six to Deep Blue, waving his hand dismissively as if to warn humanity *It's over*, we've come to accept the superiority of machines: they can outthink us, outplay us, and outlearn us. But in Xavier Snelgrove's words, they think like machines, not humans, and to quote Dave Ferrucci, head of the team that developed Watson: "They lack the human experience to put the information into context."[32]

We could accept that we are all losers in the ongoing competition between human brains and computers: let's just abandon hope and let the AIs tell us what to do. But Kasparov didn't do that. Shortly after his defeat by Deep Blue, he went on to create a new version of chess, played by human/machine teams known as "centaurs," named for the mythical beast that was half man, half horse. In 2010, he wrote:

> In chess, as in so many things, what computers are good at is where humans are weak, and vice versa. This gave me an idea for an experiment. What if instead of human versus machine we played as partners? My brainchild saw the light of day in a match in 1998 in León, Spain, and we called it "Advanced Chess." Each player had a PC at hand running the chess software of his choice during the game. The idea was to create the highest level of chess ever played, a synthesis of the best of man and machine.[33]

Kasparov pointed out that anyone can now buy a chess program for $50 that would run on a PC and beat a grand master every time, something that's proving useful in developing a new generation of high-level players, especially in places where it is difficult to find equally skilled opponents.

I wonder whether Kasparov's approach shows a more hopeful way to look at AI than simply as an inhuman jobs-killer. Poised at the end of the second decade of the twenty-first century, we could be looking at a way of solving the problem of our perpetually-on work cycles. Remember, we humans are the ones who set the context and content for deep learning; the machines don't know anything we don't know, because they receive their insights and information from us—from the interiors of rooms, to the faces of cats, to Atari games. Perhaps, instead of pushing us into irrelevancy, the machines actually have our backs; maybe AIs will augment our intelligence and carry some of the load of our 24/7 work lives, instead of replacing us as the dominant species. AIs could be about to help us survive and maybe even thrive in what has become an increasingly uncertain, fast-moving world. Maybe becoming a centaur—a human/machine team, navigating the world together as partners—is the next inevitable step for both the robots and us.

I hope so. Because otherwise, in the immortal words of Bill Paxton's character "Hudson" in *Aliens*: "Game over, man. Game over."

Chapter 5

HITCHING A RIDE IN A DRIVERLESS CAR

2025

We decided it was purely a matter of artificial intelligence.
All we had to do was put a computer inside the car,
give it the appropriate eyes and ears, and make it smart.

Sebastian Thrun, *Scientist*

I found the key to the universe in an old parked car.

Bruce Springsteen, *Rock Star*

I booked our ride for 9 a.m. At precisely 8:59:59 a.m., I watch a white Volvo sedan, model year 2025, stop in front of our house.

"Ron! Car's here!"

"Yeah, yeah," says Ron, carrying our bags out to the curb.

The Volvo's trunk hatch lifts with a sigh. As we load our luggage, I realize the sigh did not come from the car, but from Ron. The driver and passenger side doors open soundlessly and we slide in.

Inside the car, Frank Sinatra and Antonio Carlos Jobim are softly singing "The Girl from Ipanema," a duet recorded in 1967 that we have on our house playlist. A voice says: "Welcome, Terri and Ron. Please review our planned route today from Toronto, Canada, to Tenafly, New Jersey, on my dashboard screen. Estimated travel time 8 hours,

6 minutes, 54 seconds. If you want to program a stop along the way or an alternate route, say, "Reroute." Otherwise we'll be on our way in approximately two minutes."

"What if we need a coffee or a bio-break?" I ask. The car's precisely timed trip plan worries me. "Can't we be . . . you know . . . *spontaneous?*"

The car pauses for a fraction of a second. As if it's thinking. Probably processing that word, "spontaneous." Most of the time, it's the last thing you want your self-driving car to be.

"Naturally," says the car. "Simply alert me if you would like to stop your journey for water, snacks, or to relieve yourselves."

"How do we do that?" asks Ron.

"Oh, you can just say something like, 'I need a break,'' the car's voice explains. "Or words to that effect. Some advance warning would be appreciated but weather and traffic permitting, I can also be . . ." The car pauses a beat: "*Spontaneous.*"

"I think I've taught the car a new word," I murmur to Ron in a low voice, as if it can't hear me. "These new models learn, you know."

"Car, reroute," commands Ron.

"Certainly. Here are the options." On a dashboard screen, a map of alternate routes appears. Ron touches a green one marked SCENIC.

"We'd like to see the Catskills," he says. "Maybe even stay overnight, and finish the trip in the morning. Can we do that?"

The car pauses. "I am already reserved for a guest arriving at Newark Airport for 8 a.m. tomorrow. Would you like me to book an alternate car for you?"

Ron and the car negotiate, setting up a two-day trip to Tenafly instead of the one-day trip we'd originally planned. It's not a bad idea: our friends in New Jersey aren't arriving home from a trip of their own until tomorrow, so we'd be staying in an empty house. Why not take the time to do a little hiking?

"I hear Buttermilk Falls is beautiful," says the car.

"How do you know that?" I ask.

"I took another rider there," says the car. "And it's four out five stars on the Lonely Planet Guide for New York State. Shall I book you a cabin?"

While the car makes our travel arrangements, Ron revises the route on its touch screen. As they go back and forth about secondary highways, timing, and the alternate car that will pick us up at the Buttermilk Falls Lodge, I realize that the car's voice sounds familiar. It could be one of our sons talking to us. I suspect this is no coincidence but a way to make us feel comfortable. As if we were on a road trip with one of the kids.

This algorithm is good. *Very* good.

I try to relax. I've been worried about this trip, our first with a fully autonomous, shared vehicle. We've already discovered that short-hop car-sharing can be smelly if the previous occupants were over-perfumed or under-bathed. Some of the car sharing companies don't clean up between passengers; you don't even want to know the types of things I've seen in the grubby backseats of some shared cars. Dirty diapers. Doggy bags (used). Empty soda cans. Condoms. Apple cores. Cleaning up between rides is one of the biggest challenges faced by both the car companies and the people who use them, especially for daily commuting.

So, for this eight-hour drive, we decided to pay for a higher-end vehicle. I'm not disappointed—the Volvo is spotless and it's even retained that "new car smell." It comes with a cooler full of drinks and snacks, specifically selected to reflect our online shopping preferences. But Ron isn't so sure. Next to me, in the driver's seat, he instinctively puts his hands on the wheel.

"I've got this, Ron," says the car. "You're welcome to keep your hands on the wheel and direct the ride, of course, but for optimal performance, I recommend hands-free. I'll let you know if your attention is required."

Ron hesitates, lifts his hands from the wheel, and drops them to his lap. "Okay, thanks."

"You're welcome," says the car.

We're free now to look out the window, read, sleep, or watch movies on the dashboard touch screen. I swipe my way through the available content: hundreds of movies, TV shows, YouTube channels, you name it.

"This is like being on a plane," I comment.

"I've heard that one before," says the car with a chuckle. "I can assure you that I'll keep all four wheels on the road."

Ron groans. "You can skip the jokes."

"My apologies," says the car. "Would you like me to stop talking altogether or alert you when we're approaching rest stops and dining areas?"

"Just shut the . . ." Ron starts to say, but I talk over him.

"Yes, please, do alert us to upcoming rest stops and restaurants."

"Just don't try to sell us anything," warns Ron.

Now it's my turn to sigh. This is actually one of my favorite parts of autonomous riding: duty-free flash sales at factory stores and outlet malls. New Jersey is famous for them.

"We could use a new couch," I point out.

"How are you planning to carry it?" asks Ron.

"We'll have it delivered. By the time we get home, it'll be in our living room."

I settle into my seat to flip through the infotainment on the dashboard touch screen. Meanwhile, the car plays another old favorite: Bruce Springsteen's "Thunder Road." The lyrics are all about a girl, a boy, and a car that takes them far away from a "town full of losers" to a new life together. That's been a big theme in American rock music since its inception in the 1950s. There has always been a sense of freedom and romance to the open road that's starting to vanish like a slow car sinking into the horizon in a rearview mirror.

I can barely remember the traffic jams and horrific accidents of the 2010s. Even during heavy commuting times, traffic in 2025 moves smoothly, each car just a couple of feet behind the other.

Some drivers still own cars, but younger ones overwhelmingly prefer car sharing. Most are not even bothering with a driver's license: as autonomous vehicles prove themselves in the real world, insurance companies are giving up the requirement for a licensed "safety driver" at the wheel. Trucks travel on reserved commercial roads in convoys where humans are no longer at the wheel, or only onboard to monitor the cargo.

We're rolling along in a gridlock-free platoon, each car just a few feet behind or ahead of the others, using data from Lidar, radar, and sensors to maintain an even distance. In a sense, data allows the cars to "see" their surroundings. If one slows down because of weather, a deer leaping onto the road, or a child chasing a ball, all the other cars "see" and react too.

When cars are shared, they're on the road 24/7, unlike in 2018, when cars were parked 96 percent of the time. Most parking lots and garages have been demolished because they're just not needed anymore. Cities are starting to get greener, as driveways are transformed into lawns, and parking lots are turned into parkland.

Of course, you could *buy* a self-driving car, but even owners don't let their vehicles sit idle. Instead they use an Uber-style app to rent them out to others. There are even niche car-sharing services specifically designed for children or passengers with disabilities.

Except for a wave of nausea from motion sickness I experienced during a movie, our trip was eventful, safe, and pleasant. And the car was right: Buttermilk Mountain Lodge was lovely.

Was the old myth of the freedom of the road ever really true?

Let's leave Ron and I riding to New Jersey in 2025, and roll back the clock seven years to 2018. We hear about breakthroughs in autonomous vehicles just about every day. More and more driver-assistance features are being added to new models all the time, offering some idea of what full autonomy would be like. And yet, most of us have never ridden in—or even *seen*—a self-driving car. The rollout of these cars is starting to feel like the slowest moon shot in history. The biggest hurdle may not be technological, legal, or insurance related. The biggest roadblock may turn out to be *us*.

Our cars are more than transportation: they are offices, kitchens, and party rooms. Confessionals. Places of seduction. Private spaces, as personal as their owners' homes. And they're trash cans, judging by the backseats of most commuters. They're getaway vehicles, whether from a jealous lover, a nagging parent, or the scene of a crime. They're a wailing baby in a car seat, circling the block at 2 a.m. with an exhausted parent at the wheel. Try to imagine Cherry of Springsteen's "Born To Run" strapping her hands across the engines of an autonomous motorcycle. Or belting out Tom Cochrane's "Life Is A Highway" while your Google X Car putt-putts you from Point A to Point B, spits you out at your destination, and earnestly beetles away to pick up another car-free passenger. For hands-free autonomy to work, our love affair with driving is going to have to change.

There are lots of reasons to hand our keys to robot cars. Like Gort, Robby, and other pop cult robots that became popular during the nuclear paranoia of the 1950s, the driverless car could save us from our own worst impulses. Despite a plethora of semiautonomous safety features on new cars—collision avoidance systems, lane departure alerts, and more—traffic accidents are *rising*: forty thousand Americans die in car crashes annually, the equivalent of a hundred jumbo jets falling from the sky each year. Eighty percent of these crashes are caused by alcohol, speeding, and distracted driving—*human* errors, not equipment failures.[1] By 2040, when every car on the road is fully autonomous, road accidents are expected to drop by 80 percent.[2]

And then there are all the other problems caused by cars: gridlock, air pollution, stressful commutes, the urban blight of parking lots and garages, and skyrocketing insurance claims as we continue running into one another while texting home that traffic is fierce and we'll be late for dinner. All because of an expensive machine that depreciates the moment you drive it off the lot, and sits idle for 96 percent of its lifetime.[3]

And yet, few Americans *like* the idea of giving up the wheel to that of a driverless car or believe it'll be any safer: our faith in self-driving technology actually went *down* between 2016 and 2017,[4] probably due to a number of well-publicized accidents. According to the annual Allianz Travel Insurance Vacation Confidence Index, "people feel safer with the idea of space travel, supersonic travel, and flying cars than they do with autonomous vehicles" and only 22 percent of Americans are "very interested" in self-driving cars. Of those who are "not interested," 65 percent give safety concerns as the reason.[5]

Think about it: most of us feel safer flying to Mars than riding in a driverless car. At least there'd be lots of room to park.

My press pass to the 2017 TU Automotive Detroit Conference gives me a chance to talk to autonomous automakers and their suppliers in their natural habitat, something that's proven surprisingly difficult for me. Even my alma mater, a university heavy on science and engineering, and with a whole department devoted to driverless car R&D, wouldn't "open the kimono" for me, as Steve Jobs put it when he learned the secrets of Xerox PARC in 1979. At first I thought I was being snubbed because I'm writing about the touchy-feely topic of human-robot relationships rather than the hard sciences of data,

2017 Lexus IS Ad: "Enjoy the thrill of driving while you still can."

How do I explain this? It was exhilarating . . . nimble . . . responding to my every touch . . . moving faster than the wind . . . that feeling of pure . . . driving. *It was amazing!*

So runs a 2017 ad for the luxury Lexus IS, wistfully voiced by an older man in the near future, who struggles to describe the thrill of driving in the good old days before he was relegated to the backseat of an autonomous vehicle. It's like listening to someone describe the first time they had sex. Sitting in his self-driving car, his internal monologue sounds like he's mourning his lost virility.

It seems that my generation, the aging baby boomers, won't willingly abandon the steering wheel until someone pries it from our cold, dead hands.

Lidars, and sensors. So I was relieved to learn that I'm not the only one being detoured off the driverless information highway: in researching their book *Driverless: Intelligent Cars and the Road Ahead*, Hod Lipson, a mechanical engineering professor at Columbia University, and tech journalist Melba Kurman, faced a similar challenge:

> Our first lesson was that players in the autonomous vehicle industry keep their cards close to their chests. Of the employees we contacted at half a dozen car companies, not a single one responded to our email requests for an interview. When we reached out to Google X, the division of the company developing the driverless car, after several repeated queries, an administrative assistant politely pointed us to an exhibit on the history of driverless cars . . .[6]

Ouch. Driverless car R&D sounds like *Fight Club;* the first rule is don't talk about it, and the second rule is . . . well, you know. That's why my automotive engineer friend Jane suggested I join her in Detroit at the world's biggest conference for connected and autonomous cars. With 3,500 attendees representing 150 companies from over thirty countries, everyone who is anyone in automotive autonomy is here. (Actually that turns out to be not *quite* true—there are three interesting absences. More on that later.)

Ironically, the drive to the conference would provide a lesson in why autonomous cars make sense.

I'm riding shotgun in a rented Toyota Corolla with Jane at the wheel. We're one hundred miles outside of Detroit in heavy traffic. As usual, Jane is multitasking at light speed, driving one-handed while peering at the navigation app on her iPhone and chatting to me nonstop. Jane is a whip smart, high-twitch muscle fiber, type A personality who usually instills me with complete confidence. But right now, she's making me nervous.

I diplomatically suggest that I take over the navigation responsibilities. No sooner do I remove the phone from her fingers than she starts rummaging in her purse, replacing the app with an apple. "Better eat this before we get to the border."

I despair of getting Jane to focus on highway driving (and nothing else), but she's right about the forbidden fruit, one of those rules that get to be top of mind when you cross the border a lot. She finishes her contraband snack and tosses the core out the window. It's biodegradable, after all.

At the Michigan border, we show our passports and answer the usual questions. Nationality? Canadian. City of residence? Toronto. Any fruits or vegetables in the car? Not anymore. Reason for our trip?

"We're going to a conference," Jane explains.

The guard looks disinterested. "Okay, then. Have a nice day."

A nice day *was* our plan, until we hit hellacious rush hour traffic in the complex tangle of freeways and interchanges in Detroit's outer suburbs, just as Jane's phone starts running out of juice. With the iPhone's battery dying, I start scribbling directions by hand. Then I notice something weird about our final destination.

"The app is sending us to Temple Shir Shalom in Farmington Hills," I tell Jane.

She frowns. "Is that a synagogue? What happened to the Holiday Inn Express?"

I shrug. "Maybe the app thinks we need divine intervention."

"Shit," states Jane, and commandeers the iPhone again to try to reset the directions one-handed at ninety miles per hour. I break into a light sweat.

"Let me do that," I urge, then, not to hurt Jane's feelings, I point out: "You don't have your reading glasses on."

Reluctantly, Jane surrenders the phone and I finally figure out how to reset the app just before it dies in my hands.

It's times like these when I think we can't get driverless cars fast enough.

Despite consumer skepticism, there's a good chance that the first personal robot you own will be a driverless car. Prototypes of autonomous vehicles can already do everything R2D2 could do: they can see what's around them, using cameras, sensors, and Lidar mapping technology. They can use artificial intelligence to identify stationary and moving objects, like walls, buildings, bicycles, and pedestrians, and anticipate what they will and won't do. They can keep an eye on you for signs of fatigue or frustration, like yawning or pounding the steering wheel, and react appropriately by waking you up, or suggesting you stop for a coffee or a rest break. They can carry on conversations with you and your passengers in natural language. And, like the games-playing AIs in the previous chapter, they can learn, both from their own experiences and from other cars. They also offer a glimpse of how we'll interact with robots in our homes and workplaces one day. Think of the driverless car as the synthetic canary in the automated coal mine of human-machine relationships.

So when does all this start, for real? Most carmakers have settled on 2025 as the year that driverless cars will likely start rolling off production lines. Honda, for example, has announced that's the year it will start producing cars with stage 4 autonomy, when the car can do *all* the driving, but a licensed driver still needs to be prepared to take the wheel should something so unpredictable happen that the car can't figure out what to do, such as a freak snowstorm covering all the car's sensors, or a UFO landing in the middle of the highway. Drivers would become like airline pilots: not strictly necessary, most of the time, but critical some of the time. Although planes can fly on their own, there are still unpredictable situations when human perception and experience are needed: that's why trained pilots are still in the cockpit. Not until we reach stage 5 autonomy—no steering wheel or pedals—will we be completely free of the need for driver education, licensing, traffic signals, signage, and (possibly) insurance. As for car sharing, a new study from the Center for Automotive Research in Ann Arbor, Michigan, finds that private car ownership will still be the norm in the next decade, although car sharing will increase to 30 percent of the market, a big jump from the 4 percent of rides it represents today.[7]

With so many people using the same cars, car-sharing companies will face the challenge of having to clean up between rides—the left-behind apple cores, coffee cups, and vomit—carsickness, being a side effect of riding in a driverless car for up to 22 percent of drivers, rising to 37 percent when they try reading or writing in the car.[8] Not to mention the question of who's responsible: no matter how much safer autonomous vehicles will be than human-driven cars, there *will* still be accidents, caused by things like weather, equipment malfunctions, and hacking. And of course in 2025, driverless cars will still be sharing the road with nonautonomous and semiautonomous old beaters from the 2018 to 2024 model years. That could mean different roads for different car types, if we can rebuild our highways quickly enough. But eight years isn't much time for a complete overhaul of infrastructure, especially in an old manufacturing city like Detroit, which, although going through something of a renaissance, still struggles to find the money to keep the lights on and the potholes filled. Throw in the unpredictable, non-roboticized behaviors of pedestrians, cyclists, and the odd leaping deer, and the coming years are going to be interesting for road warriors.

Driverless car R&D falls into two opposing camps: one is the Google model, which treats autonomous driving primarily as a software problem, with the "hardware" of the car a secondary consideration.[9] (This may explain why Google cars to date look like shapeless blobs.) Traditional car manufacturers, on the other hand, are taking a go-slow approach, evolving their vehicles into fully autonomous cars by adding more and more sensors and software that provide Automated Driver Assistance Systems (ADAS)—collision avoidance, cruise control, automated parking, and so on—until we gradually take our hands off the wheel without giving it a second thought.

But this could all suddenly change if one company comes up with what's known as a "killer app"—the type of thing that happened when Apple upended the music industry with iTunes. If that happens, the rules of the self-driving car game could change very quickly, with one company zooming out in front and everyone else scrambling to catch up.

Despite how futuristic driverless cars may seem, the concept of autonomous driving has been around for almost eighty years, although the original idea was to automate roads, rather than cars themselves. At the 1939 New York City World's Fair, millions of visitors took in "The Futurama," a miniaturized model of a look-ma-no-hands city that guided cars on and off highways using radio waves. GM predicted that automated highways would be the norm by 1960. The idea was shelved during World War II and resurrected in 1957, when GM partnered with RCA to design an electromagnetic highway. But it never got past the prototype stage. In the real world, automated roads turned out to be wildly expensive and impractical.[10]

During the early years of the Apollo space missions, the moon rover was initially conceived as a self-driving vehicle, but in the end, NASA nixed the idea: it was important, they said, to have brave astronauts at the wheel, both for public relations and to make the crew feel that they were in control.[11]

Then came the twenty-first century, 9/11, and the war in Iraq. To save soldiers' lives, Congress directed the Pentagon to develop unmanned vehicles.[12] In 2004, the Defense Advanced Research Projects Agency (DARPA), the same branch of the US military that funded "artificial human" Shakey in the 1970s, offered a $1 million prize to the winner of a 120-mile driverless car race through the Mojave desert. The DARPA Grand Challenge was intended not only to get university and corporate research teams excited about driverless cars, but also hobbyists—the same breed of tinkerers as the old Home Brew Computer Club whose creativity helped transform the first minicomputers into PCs. Out of fifteen competitors in that first race, not a single car made it further than seven miles before crashing, flipping, spinning in circles, or breaking down.

In the aftermath of the challenge race, DARPA doubled the prize to $2 million. Within a year, more than one thousand people were working on developing a driverless car. That's a lot of brainpower applied to one problem. (Interesting side note: one of the failed competitors in the 2004 race, David Hall, whose past experience was manufacturing subwoofers for home theater systems, never competed again, but went on to the develop the Lidar technology still used in self-driving cars to map terrain.[13]) In 2005, out of forty-three semi-finalists who competed on a speedway track that simulated some of the biggest challenges in the desert course, twenty-three completed at least one test run—a big leap in just one year. In the final heat through the Mojave, the winner was an autonomous VW Toureg nicknamed "Stanley," built by a Stanford University team led by AI expert Sebastian Thrun. He went on to join Google, which announced its own self-driving car project in 2009. Tesla followed Google down the road to autonomy in 2014 with its semiautonomous autopilot feature, and the promise of fully autonomous electric cars in the future.[14] Then the traditional car manufacturers started getting on board by introducing more and more semiautonomous features and buying up car-sharing services. In 2017, Apple applied for a license to test self-driving cars in California, but the development of an "iCar" has been kept strictly under a Cone of Silence.[15] That brings us to where we are today, a transition time of prototypes, promises, and ads that play on our premature nostalgia for "real driving."

In the parking lot of the conference center, a demonstration area has been set up, fenced in by signs reading AUTONOMOUS TEST TRACK. My priority over the next two days isn't just to learn about driverless cars, but to ride in one—or more correctly, *hitch* a ride in one, something I'd missed out on when I was trying to catch a self-driving Uber in Pittsburgh. I'm a nondriver, but I did pass my driver's test, mostly so that I'd have proof of age at the liquor store. I haven't driven in over thirty years (or needed proof of age, come to that).

Inside the convention center, it feels as if it is *already* 2025: the focus is on the future, when cars will be as much computer as vehicle. This

Every car chase in the movies versus Johnny Cab (*Total Recall*, 1990)

In the great car chase movies—*The Fast and the Furious, The Blues Brothers, Bullit, The French Connection, The Italian Job, Mad Max*—the driver is always the hero (or more often, antihero).

In the sixties, seventies, and eighties, a few movies and TV shows featured fantasy cars that could think, speak, browbeat family members, fight crime, and murder teenaged bullies. *My Mother The Car* was a 1965 situation comedy in which the car owner's dead mother was reincarnated as a vintage automobile. In 1982 we saw both Stephen King's sentient, homicidal Plymouth Fury *Christine* and K.I.T.T. the intelligent, crime-fighting Pontiac Firebird in the TV show *Knight Rider*. With the possible exception of K.I.T.T.—who was still very much the sidekick of David Hasselhoff's hunky character, Michael Knight—none of these semi-robotic cars could be called heroic (although Christine was a great horror movie villain).

Even one of the more memorable self-driving cars in the movies—*Total Recall*'s Johnny Cab—is portrayed as being driven by a leering marionette. Johnny responds to Arnold Schwarzenegger's command to "Drive! Drive!" with a maddeningly chipper, "Can you repeat the destination?" Some getaway car! You can't blame Arnie for ripping Johnny out of the cab and the taking the driving into his own hands.

I'm still waiting for stories starring cars that fall in love, lose their tempers, and turn into fully realized heroes and villains. (Okay, besides Disney's animated *Cars* franchise.) One film that comes close is *Logan* (2017). Set in 2029, the Marvel comic book character, Wolverine, is an aging antihero who makes his living as the driver of an autonomous limousine. The automotive "bad guys" are platoons of driverless transport trucks. Without front cabs, they look like hulking, headless monster robots, rolling along long, lonely highways that are devoid of humanity.[16]

is no automotive show full of spokesmodels in bikinis draped over concept cars, but a networking conference for industry heavy hitters and automotive journalists. Despite a bad case of imposter syndrome—I don't even drive, for god's sake—I fuel myself with the complimentary coffee and croissants before heading to a series of headline talks and "fireside chats" that will set the tone for the next two days. I'm surprised when the speakers who kick things off aren't carmakers, but tech companies focused on a magic word that promises a way to both "power" autonomous vehicles and profit from them: *data.* That's what really controls the self-driving car: data, to allow it to see, talk, move safely, learn from experience, and manage the people riding inside it. Data is as much of a fuel for autonomous vehicles as gasoline or electricity.

At first, it's all very gung ho stuff—lots of lofty ideas

that are big on jargon but short on specifics. We're undergoing "the largest change in the transportation industry since Ford's Model T," with 26 billion devices connected to the Internet (and one another) by 2023, many of them cars. "Every product is a service waiting to happen." "The speed of change will never be as slow as it is today." Yadda, yadda. I yawn and check my email, worrying that the speakers are not going to tell me something useful. And then Intel takes the stage.

I admit, I have a mild corporate crush on Intel. Not only is it the company that gave us the microchips that ushered in the era of desktop computers, Intel was cofounded by Gordon Moore, whose famous Law predicted our era of ever-accelerating technological change. I love those old "Intel Inside" ads from the 1990s, with bunny-suited techs boogying as they put Pentium chips into laptops to the *bee-bong-bee-bong* signature sound, which ranks just behind the Apple log-in double C major chord chime as my favorite tech sound.

In a talk called "Autonomous: Why Not Sooner?" Doug Davis, Senior Vice President and General Manager of Intel's Automated Driving Group, gives examples of technologies that took decades to gain traction: seat belts were not a standard feature in cars until 1968. Air bags were invented in the 1950s, but weren't required until 1998. Antilock Braking Systems, better known as ABS, took sixty years to get into cars. Automatic transmissions took a staggering forty years to gain acceptance. Davis' message seems to be that, even if it will save lives, it takes a long time for automotive technology to be perfected, approved, implemented—and to get people to buy in. This is not an industry that disrupts, but one that evolves . . . *slowly*.

Davis draws a parallel between the introduction of driverless cars and personal computers. PCs had already been around for ten years when IBM launched its first desktop computer. Designed primarily with off-the-shelf parts, the IBM PC was the game changer that finally persuaded people that they couldn't survive without a computer on their desks, a notion that would have seemed absurd just a few years earlier. Driven in part by IBM's rock-solid brand and extensive dealer network, the IBM PC became the killer app of its day—the product that set the benchmark for every personal computer that followed. Whichever driverless car captures driver confidence and enthusiasm first will become the IBM PC of autonomous mobility.

The Intel presentation wraps up with an ad called "Letting Go," a collage of images of people holding onto something for dear life: a little boy clings to the edge of a pool while an indistinct figure—a parent or swim instructor—treads water nearby; a tired boxer hangs onto the ropes, trying to get up his nerve and return to the fight; a terrified bungee jumper clings to a bridge; a trick cyclist grips the handlebars of his bike; a physio patient braces himself on handrails as he tries to walk. As the music rises to a crescendo, the boy lets go and swims, the boxer takes a swing at his opponent, the bungee jumper screams and leaps into the void, the stunt cyclist stands on his saddle, and the patient takes a few independent steps. Cut to a man lifting his hands off a steering wheel. The copy line on-screen reads: *Amazing things happen when you let go.*

With all due respect to Intel's ad agency, the people in the ad aren't letting go, but taking control. Self-driving cars require that we *give up* control and accept the role of passive passengers. That's a big communications challenge for an industry built on the promise of personal freedom and convenience.

But remember, Intel's customers are carmakers, not drivers. Intel no doubt wants to be the chip of choice for driverless cars, just as they are for computers. Could their brand help build our confidence that robot cars can drive better than we can?

A line of small type, likely added by a nervous marketing lawyer, appears as the hands hover off the steering wheel: *Do not attempt. Autonomous cars not yet available.*

That legal boilerplate summarizes the overarching question for people like me. *If not now, when?*

An answer, of sorts, comes in the next session, a fireside chat between two German CEOs: Thilo Koslowski of Porsche Digital and Holger Weiss of German Auto Labs.

"It's done, check the box," says Weiss confidently in his crisp accent.

He's referring to autonomy, as if the driverless car was a done deal: we've done it, it's ready to go, let's move on. *Really?*

I'm startled by his confidence but what he actually meant is that the technology is already well developed. Not perfect, perhaps, but far down the road to full autonomy and something called connectivity.[17] In 2018, new cars are not only equipped to connect you to the Internet, but with vehicle-to-vehicle (VRV) systems that let your car talk to other connected cars ("Hey watch out, slippery road ahead") and vehicle-to-anything (V2X) systems that link your car to the Internet of Things—your home, your office, your phone, and even fast-food restaurants at highway stops ("There's a McDonald's in fifty miles and by the way here are two McCafe coupons for you and your friend"). No need to squint at smartphones: these are "digital cockpit solutions" that travel with you like know-it-all friends.[18] Yet, in 2017, most drivers chose *not* to activate these systems. In the United States, fewer than 15 percent of the vehicles on the road last year were connected.

Why the slow uptake? Simple: people have to pay for it. Jane snorts at the idea of charging drivers extra for data. "People are already connected through their phones," she points out. "They're using Facebook for free. They're not going to pay extra for a service they already have. They'll just get in their cars and check their phones."

With that in mind, we enter the trade show where mighty IBM shares floor space with start-ups from Finland, Israel, France, and just about everywhere else in the world. Before Jane and I part ways to follow separate tours of the show—Jane with the engineers and software designers who want to know about the serious tech stuff, me with the group interested in the touchy-feely "customer experience"—Jane gives me some background. Everyone here (including Jane) wants to network with "OEMs" (original equipment manufacturers), also known as "Tier Ones" (GM, Ford, Fiat-Chrysler, VW, Porsche, and other automotive manufacturers). "Tier Twos" are the regional offices of Tiers Ones. Tier Threes are car dealerships, the group with the most to lose in the years ahead as car sharing grows. And then there are the start-ups, circling the sluggish Tier Ones, Twos, and Threes like perky satellites. The car sharing companies are here, too. Although there's a lot of jockeying for position and what might impolitely be calling pissing contests, they all seem to more or less agree on the final destination of the journey: fully autonomous vehicles, connected to one another and "smart" city infrastructures (traffic lights, signage, other cars, even the paint on the lines of the roads), probably electric, and very likely shared.

One company invites people into a demonstration vehicle, not to drive, but to talk: they've developed a natural voice recognition system that enables the car to hear, understand, and respond to multiple voices: if you're squabbling over where to go for pizza, the car can keep up with the argument (and let you know when Little Caesar's is coming up on the left).

A company from Finland demonstrates software that provides detailed weather and road conditions in real time so drivers can adjust their driving. Eventually, this data could connect directly to the car itself. Given that the only car accident I've been in was a spinout on a slippery road in the middle of a blizzard driving home from university for the Christmas holidays with Dad at the wheel, I've always thought that the biggest challenge robots would have to face was weather. I'm comforted seeing that the system is, at that very moment, transmitting weather conditions to drivers in New Zealand.

But in the Start-Up Zone, an urban driving simulation software company gives my favorite demonstration. Given the high number of test drives that will be required for self-driving cars—literally trillions—many will have to be done by simulators. Cognata is a digital platform developed by an Israeli company that simulates typical driving behaviors in specific urban environments, based on the ways drivers and pedestrians *actually behave* in those cities. How much do people jaywalk in Manhattan? Do drivers typically run reds and fail to signal turns in Los Angeles? Do people usually go the speed limit in Seattle? What about bicycles and motorbikes in Boulder?

"Wow," I say, trying to think of the most unpredictable driving city I've ever been in. "So, you can simulate Rome? Do you ever think they'll have self-driving cars?"

One of the Cognata guys grins. "Rome? Ha—never! Too crazy."

We both laugh because we know it's true; how are you going to get the Romans off their Vespas and into driverless Fiats? And given the high level of—shall we say, passionate spontaneity in the population—will autonomous vehicles be able to adapt to the apparent chaos? Rome, like every city, has its own driving culture: watch it from a sidewalk café and you start to realize that the drivers have their own set of rules, based on certain etiquette. It's really not as chaotic as it appears. The Cognata platform is built to adapt to that.

"What country do you think will have full autonomy first?" I ask the Cognata guy. "Japan?"

"More likely, China," he answers.

He's probably right. In a 2017 study of autonomous vehicles in ten countries, China, along with Singapore and India, was identified the country with the highest customer interest in autonomous and electric vehicles.[19] The United States—where autonomous technology was pioneered—was ninth.

It's midafternoon and I'm starting to tire: there's nothing like the weariness that sets in after your fourth hour inside a windowless convention center. Jane is off somewhere networking with Tier Ones. I'm starting to think the world has turned into one big spinning Lidar, when I see what looks like a peaceful oasis in the middle of the show: a living room with couches, coffee tables, and a phone-recharging station—the Lixar Lounge. No demos, pitches, or promotional materials, just a welcoming place to sit and relax. Across the back wall of the lounge, Lixar's logo is displayed alongside those of its clients, including NASCAR.

"Self-driving racing cars—really?" I ask a rep who discreetly looks after the lounge.

No, not self-driving, she tells me. Lixar is a mobile data company that works with NASCAR to provide an immersive experience for their customers, but not to replace the drivers. I explain my interest isn't just in self-driving cars, but robots in general. The rep tells me that their company's lead data scientist compares the experience of humans adjusting to robots to the way people first reacted to elevators.

I smile, thinking of Dad's elevator to nowhere. When elevators were introduced in the nineteenth century, people were leery of finding themselves hanging by a cord in the shaft of a multistory building. What would stop it from snapping and plunging passengers to their deaths?

As skyscrapers grew higher, riding in elevators became a necessity. Trained operators manually moved the cars from floor to floor, instilling a certain amount of confidence in their passengers: a human was clearly in control. But when push-button models came into common use in the 1950s, the disappearance of the human operator caused discomfort and fear in passengers, as well as the spooky sense that the elevator had a mind of its own: how could simply pushing a button "tell" the elevator where to stop, when previously it had to be maneuvered between floors?

You could think of the elevator as a fixed-in-place robot, like a factory robot, or a self-driving car on a vertical track.[20] The level of trust required to ride in an elevator foreshadows the emotional process we'll go through to accept driverless cars.

Phone charged, mind relaxed, I head to a panel that speaks to one of the big questions about self-driving cars: customer acceptance. The session kicks off with the moderator asking for a show of hands: how many people would put a child or elderly person into an autonomous vehicle?

Only a smattering of hands goes up—and this is an audience of *industry insiders*. Their response reflects public opinion: three-quarters of Americans wouldn't trust a child or an elderly person in an autonomous car.

Speaker Kristin Kolodge, Executive Director of Driver Interaction and Human Machine Interface for the global market research company, J. D. Power, says: "It comes down to emotional trust." She goes on to explain that studies show that drivers are "overwhelmingly thrilled" with the lower levels of automation provided through Automated Driver Assistance Systems. But how do you build trust in *fully* automated driving?

Everyone on the panel agrees that it's a matter of getting people comfortable with what's coming in the future. Standardizing terminology would help. For example, ADAS features vary in name from one car make and model to another (and don't always work the same way) creating what Kolodge calls an "alphabet soup" of options. The dizzying number of names and acronyms makes the cars hard to figure out for the average person. What can the car do? What can and should the driver be ready to do? The problem is big enough that the National Safety Council runs an education program called, "My Car Does *What?*"

The moderator wonders whether it's a psychological challenge—a feeling of loss of control. How is it possible to change that?

Alex Epstein of the National Safety Council weighs in with something I've been thinking about all day: changing the culture of driving.

"Cars have been marketed as keys to the open road. Sexy and sleek. Now they have to be sold as 'mobility.' And people are misled about the year full autonomy is coming. Carmakers have to drive trillions of miles before we'll have anything truly autonomous."

One of the panelists theorizes that younger drivers of Generations Y and Z "don't have the same desire to be race drivers as their parents. They'd rather enjoy the infotainment system."

Research shows that Generations Y and Z—born in the 1980s and later—are more open to self-driving cars than the pre-boomers, boomers, and Generation X. Maybe it will take younger drivers showing older ones that they trust autonomous vehicles in order to make that shift

The check box for driverless cars may be "ticked," but the challenge of winning hearts and minds hasn't even really begun.

By 5 p.m. I still haven't had my ride in a driverless car and I'm getting texts from Jane, *Where are you?*

Heading for the pre-networking cocktail party at the Lixar Lounge.

The what?

It looks like a living room. Trust me, it's the best place at the conference to relax.

The Lixarians aren't just brilliant data scientists, they're smart marketers. By setting up a comfortable, stress-free, congenial place to hang out without being constantly pitched, they've attracted OEMs, start-ups, and for all I know Tier Twos and Threes. (Okay, granted, at the end of a long afternoon, their lounge also hosts a complimentary bar.)

We're still hanging out when Jane appears. Of course, she and one of the Lixarians recognize each another from a past event, and start catching up on one another's careers and kids. It might be the biggest autonomous car convention in the world, but the industry feels small all of a sudden.

When I explain the premise of *Generation Robot,* we fall into the inevitable discussion of just how much robots will take over our lives—and more precisely, our jobs.

"So what's the solution?" I ask.

"Study math," suggests a Lixarian. Jane agrees.

I sip my drink and don't point out that AI is already *really, really* good at math. And that there are some predictions that the future is not in building and designing robots, because ultimately they will be able to do that themselves. Car factories are a good example of that. What will be left for humans to do?

It seems to me that the two things that haven't been roboticized are imagination and ambition. Robots can think but, so far, they can't dream. It's the human ability to do what I'm doing with Jane, and the Lixarians that's so uniquely human—exchanging ideas and connecting. It's why the Lixar Lounge was set up as a place to get to know one another. It's a setting that could incubate really cool ideas, like the cocktail party that led to the invention of UNIMATE. Wozniak and Jobs were inspired by Atari games and a ragtag bunch of hippie geeks in the Homebrew Computer Club. A guy working for Intel, asked to make a chip for a pocket calculator, went rogue and designed one that ushered in the era of desktop computing. Stuff happens because humans conceptualize and dream big. That's something no roboticist or AI scientist has been able to automate.

The networking party is closing for the night. Tomorrow, I am determined that I will take my first ride in a self-driving car.

Day Two. I head to the conference early, arriving before most of the attendees have slept off their networking cocktails. I want my ride into the future.

The Perrone Robotics guys (yes, all guys) are just getting the autonomous test track ready; so far, two men in suits are chatting as they wait their turn. I ask one of the technicians if I can join the two men already in line, adding that I'm writing a book about human-robot relationships.

"Well, what's a robot?" he asks. "A dishwasher is a robot, but we don't think of it that way—eventually people won't think of self-driving cars as robots either."

He's right—the "what's a robot" debate rages on—although a roboticist wouldn't think of a kitchen appliance as a robot unless it was AI enabled, connected with the other systems in the kitchen, and mobile, moving from table- to countertop, gathering dirty dishes on its own.

Finally, the car is ready. "We'll get going in just a few minutes, ma'am," a bearded guy in a ball cap tells me, charming me with his southern manners. He explains that the driverless car is a Lincoln Continental MK2 hybrid, equipped with $60,000 in sensors, radar, and Lidar. Nicknamed "Max," it has been trailered up to Michigan from Virginia, one of the few states where autonomous vehicles can travel the roads freely. Here in Michigan it's stuck on the test track with a driver at the wheel. California and three other states have specific "driverless licences"—hard to obtain—and the rest of the country is a hodgepodge of regulations. But this is changing: the US Senate recently passed a bill to standardize laws for autonomous vehicles throughout the country.

Max keeps us up to date on what he's doing. *Hello, my name is Max. I'm doing a pre-check.* A moment later, he reassures us: *Pre-check okay.*

And we're off. When we drive up behind a parked van set up on the track as an obstacle, Max tells us: *I'm stopping for something.*

He turns on his directional signal, pulls out, and passes the van.

A young man in a white shirt and tie walks out in front of Max, feigning surprise. Max brakes. "That's our intern," chuckles the driver. The intern smiles and waves, clearly confident that Max isn't going to run him down.

Max drives a figure eight. The ride is smooth, if slow. The steering wheel turns on its own but if the driver wants to take back control of the car, all he has to do is touch it.

I'm back out in the parking lot fifteen minutes later, another group waiting to go for a spin. These industry insiders are as thrilled as I am to have a ride in a genuine self-driving car. Even for them, the novelty hasn't worn off. I feel a sense of occasion, having just ridden in the belly of a robot for the first time. I'll put this on my list of tech firsts—like the first time I used a PC or sent an email.

Despite my time in robot car land, the road to autonomy still seems foggy. I came away from the self-driving conference with contradictory answers to many of my questions. For example:

If driverless cars will save so many lives, why do we trust them so little?

We're saving lives was a constant talking point at the conference. I believe it. Even if the technology isn't perfect, an autonomous car couldn't do much worse than a human driver.

The fact is, though, that it's impossible for even the most advanced technology to work 100 percent perfectly, 100 percent of the time. So, how safe is safe enough, given that humans set the bar so low? Some analysts think that carmakers could learn from the airline industry, which has improved safety dramatically since the 1950s (although fear of flying remains one of the most common phobias).[21]

Bottom line is that standards are going to have to be set extremely high. Tesla's founder Elon Musk explained the challenge this way:

> . . . getting a machine learning system to be 99 percent correct is relatively easy, but getting it to be 99.9999 percent correct, which is where it ultimately needs to be, is vastly more difficult. One can see this with the annual machine vision competitions, where the computer will properly identify something as a dog more than 99 percent of the time, but might occasionally call it a potted plant. Making such mistakes at 70 mph would be highly problematic.[22]

Will Autonomous Driver Assistance Systems help ease people into full autonomy, or cause more crashes (because people forget that their cars aren't self-driving)?

OEMs like to say that Automated Driver Assistance Services not only ease us into the idea of the car doing more and more of the work until we're comfortable taking our hands off the wheel entirely, they keep the human in the loop, mentally and physically in the act of driving. That helps ensure that our driving skills stay sharp for the odd time we're called upon to use them. Makes sense, right?

Not so fast. Some companies (like Google) believe that going straight to stage 5 (full autonomy, no pedals, no steering wheel) makes more sense and will be safer in the long run. The reason: being disengaged from the act of driving even for short stretches of time means that the driver will have a harder time coming back into the loop. It would be safer to simply let the car make all the decisions, all the time, rather than trying to alert a driver who has fallen asleep or become absorbed in a movie. Forget about the little steps: just dive directly into full autonomy.

"The Good Wife" Season 7, Episode 7: *Driven* (November 16, 2015)

This lawyers-in-love TV series used a malfunctioning driverless car to dramatic effect in an episode about a serious traffic accident that resulted in a lawsuit. Naturally, it wasn't the driverless car's fault: an evil software developer had hacked it to run through a stoplight, in an updated twist on the classic trope of sabotaging an enemy's brakes on a winding mountain road. Despite the novelty of the situation, there's nothing new about a car hack: it's become a common way for thieves to hot-wire a vehicle's computer systems, get in, and drive away. The notion of hacking a driverless car opens the possibility of making it do your bidding from a distance, with no one at all at the wheel. IBM Watson, for one, is working on anti-hacking protection for connected cars.

The jury is out on which side is going to win this battle.

Car sharing could cut down on gridlock, but most of us don't like to share. It's hard to imagine most car owners giving up the convenience of a vehicle in the driveway, ready to take them anywhere, anytime. If the sharing economy replaces private ownership, we're not only going to need to adjust to driverless cars but to being car-less drivers—hitchhikers on the highway of life. Navigating this change in the status of the car is going to require more than sensors, Google Maps, and autonomous parking systems. It means less convenience, spontaneity, and independence.

The Big Three Carmakers were at the conference. The New Big Three weren't. At a conference with over three thousand attendees from companies like IBM, Ericsson, Intel, and all the Big Three US carmakers as well as Porsche, VW and the car-sharing giants, *where were Google, Tesla, and Apple*? Probably back home in Silicon Valley, working on their secret disrupt-the-entire-industry driverless car thing faster than anyone thought possible. Maybe I'm reading too much into their absence. Or maybe, the established brands we know and trust will not survive the driverless revolution. If cars truly become artificial intelligence on wheels, and if the sharing economy really does result in the abandonment of private car ownership except for a select few, then brand loyalty to traditional automakers could vanish.

On the other hand, the carmakers know something the tech companies don't: how to build cars. To quote an insider, a car isn't simply a washing machine on wheels. It could be that the car manufacturers will, in the end, partner with a start-up, or even with Google: it's been reported that Google may abandon its blob of a car and move toward providing OEMs with self-driving software packages instead.

As the pre-boomers, boomers, and Gen-Xers die off, their emotional attachment to the hippie-esque freedom of the open road will likely pass away too, giving Generations X, Y, and Z the opportunity to take their hands off the wheel. That generational shift matters, because no matter how amazing the technology might be, or how well prepared cities, regulators, and insurance companies become, people have to buy in to driverless cars and everything that is supposed to come with them: car sharing instead of private ownership, electric vehicles instead of gas, and data-connecting cars in a hive-like Borg that lets each car learn from every other car on the road.

2020 Tokyo Summer Olympic Games,

Tokyo has announced that it wants to become a "completely autonomous" driving city by the time the games open. The city is also revamping its digital infrastructure so that every autonomous vehicle is continuously updated about changes in driving conditions, including the presence of cyclists and pedestrians. Self-driving robot shuttles will go into service in 2019 to get ready to transport the massive crowds the Games will bring. If all goes well, driverless vehicles at the Olympics could speed up the process of bringing self-driving cars into everyday life. What better country to lead the way than robot-loving Japan, home of Astro-Boy?[23]

Cars will no longer be sleek, sexy symbols of freedom: they'll be "shared mobility solutions," more service than product, more digital than mechanical.

How will carmakers build trust in fully automated driving? It's a question that hasn't yet been answered, but here's my nonexpert opinion (borrowing from Robert Frost): the future may sound like it's right around the corner, but we have many simulated miles to go before we can take our hands off the wheel and sleep.

Chapter 6

WAKING UP THE HOUSE

2030

> Very soon, most manufactured items, from shoes to cans of soup, will contain a small sliver of dim intelligence.
>
> Kevin Kelly, senior editor, *Wired*

Before speculating about our lives twelve years from now with robot chefs, Mensa-level intelligent refrigerators, staffs of virtual servants, mind-reading shopping apps, and pushy thermostats, join me in the Wayback Machine to glimpse a home in a simpler time: 1923.

That was the year that the house where we live was built for the Reilly family: mother, father, and two daughters. The Reilly sisters never married. After their parents died, they continued living in the house into old age. The last surviving sister moved into a retirement home in 2000. That's when the place went up for sale and Ron and I took it over.

Despite the fact that the house was a sagging fixer-upper by the time we bought it, the Reillys had been well-off. Mr. Reilly was a stockbroker who had a telephone installed in the front room he used as his office—a high-tech amenity by the day's standards. Our deed shows an easement where a laneway once cut through a neighbor's property from a side street to provide a discreet back entrance for deliveries from the bread man, milkman, egg man, iceman, knife sharpener, coal man, and, no doubt, a guy with a fruit and vegetable pushcart.

The family could have ordered everything they needed with their fancy phone but they probably didn't need to. After a few visits to the Reilly home, the deliverymen would have the family's weekly purchases committed to memory and simply brought them "the usual." From time to time—knowing so much about the Reillys' tastes and habits—they would have tried to upsell them: *Hey missus, your little girls like oranges but today I got a beautiful shipment of bananas. For you, special price.*

There are a couple of reasons why the 1923 Reilly home is relevant to "smart" homes today and in the future.

First, in those glorious years before the suburbanized space age, the Reillys could get highly personalized services right in their community from a host of deliverymen coming round to serve them. The Internet is our community now. The life of the Reillys anticipated the Internet of Things, or IoT, by almost a hundred years. IoT returns us to that time when we could have all our needs looked after automatically, in that noblesse oblige *Upstairs Downstairs* way. That's kind of nice, even if we're being served by artificial intelligence instead of humans.

Secondly, like many of today's digital immigrants the Reillys were resistant to change. The egg man, bread man, and milkman retired, or died, or were put out of business by a shopping mall. The fruit and vegetable man sold his pushcart and got into real estate. But the Reillys hunkered down and tried to live their lives as they always had. For eight decades, the house remained virtually untouched. At the dawn of the twenty-first century, the basement still had a coal room, a dirt floor, and an iron giant of a boiler that pumped hot water through ancient radiators. Open the hatch at the bottom of the boiler and you could see flames dancing inside. The walls were plaster and lath. Electrical wiring was knob and tube. The Reilly house was a museum piece. Almost a time machine. That's what happens when you reject the new and cling to the comforts of obsolete technology. (It also meant that the house scared off so many potential buyers that we were able to drive down the asking price.)

That's why I've decided to stop worrying and learn to love the Internet of Things. As I would have told the Reillys, when it comes to technological change, resistance is futile. Between now and 2030, IoT will transform the way we live by making our houses so smart that they'll automatically do much of our shopping, planning, and grunt work for us. What's more, IoT won't just exist in newly built *Jetsons*-y homes, but in century homes, like ours. Which could mean yet another overhaul of the old Reilly house as new devices are installed and Wi-Fi updated. Patches will be vigorously applied, not only to the walls and roof but to Internet security.

Think of 2030 as the era of the eternal upgrade.

To predict the future, you have to start by looking at the present: where are the game-changers that will transform home life over the next twelve years? How will we be cooking, cleaning, and maintaining our homes in 2030?

The Wayback Machine

In 1996, a nonprofit organization called Wayback Machine was launched in San Francisco to create a permanent archive of the Internet by capturing all the online content that is lost when sites change or shut down. The name comes from a recurring segment on the *Rocky and Bullwinkle Show* (1959–1964)—one of the best animated TV series, ever—called *Peabody's Improbable History*, featuring a WABAC (aka Wayback) Machine that took superintelligent beagle Mr. Peabody and his friend Sherman on a zany new time travel adventure each week. Eventually, the "Wayback Machine" became shorthand for looking back in time.

Why are the Wayback Machine's archives needed? Because although most of us think that anything posted to the Web will stay there forever, most web pages have a life span of only a hundred days. Large amounts of data are permanently removed, sometimes to make our past selves look smarter than we really were. As Jill Lepore wrote in her *New Yorker* article about the Wayback Machine, "BuzzFeed deleted more than four thousand of its staff writers' early posts, apparently because, as time passed, they looked stupider and stupider. Social media, public records, junk: in the end, everything goes. Web pages don't have to be deliberately deleted to disappear. Sites hosted by corporations tend to die with their hosts. When MySpace, GeoCities, and Friendster were reconfigured or sold, millions of accounts vanished."[1]

You can browse the Wayback Machine's archive free of charge at archive.org. Their brick-and-mortar headquarters is an old church in San Francisco built in (wait for it) 1923,[2] the same year as the Reilly House.

More than one roboticist has told me that we are at least ten years from a mobile robot with the dexterity to clear a table, pick your kids' clothes off the floor, sort them for the laundry, and prepare a meal (although one stealthy start-up owned by Google might surprise us—more on them later). If that estimate is correct, most mobile household robots may still be novelties in 2030, available only to well-off families like the Reillys. In the meantime, the disembodied agents of the Internet of Things will be well entrenched in our homes, taking on many of the energy-sucking tasks in our lives. You may never have to scrub a toilet, empty the garbage, shop for groceries, or much of anything else, again.

You'll also spend zero time driving, because all cars will be fully autonomous by 2030 and private ownership will morph into corporate car sharing, at least for most of us. Private car ownership will still exist for the rich, and non-autonomous driving will be a recreational activity pursued on private roads, like horseback riding is today.

You probably won't be tapping letters on a keyboard or even a touch screen anymore, thanks to intelligent voice recognition. Desktops and laptops, already being replaced by smartphones and tablets in 2018, will surrender completely to voice-activated mobile devices by 2030. Expect the old QWERTY keyboard that came down to us from typewriters to finally be tossed into technology's delete bin.

I could live with all that, although the trade-off might be constant vigilance to protect ourselves from hackers, hustlers, and hucksters.

As I write, more than three billion devices are connected to the Internet (and potentially to you, me, and one another) and those things will keep getting smarter, thanks to our addiction to social media. Our tweets, texts, posts, YouTube videos, Instagrams, and general sharing of the minutiae of day-to-day life will continue teaching AIs all about us.

We already have smart servants like Siri, Alexa, and Google Home. Twelve years from now, we'll have a staff of them. Not since *Downton Abbey* will there be so many helpers to get us out of bed, fed, dressed, entertained, cared for, and generally cosseted. But unlike those cooks, serving girls, houseboys, housemaids, and butlers with their curtsies, bows, and behind-our-back gossip, IoT assistants will be discreet and uncomplainingly on the job 24/7. What's more, they'll never steal the silver, run off with the chauffeur, frighten the horses, or seduce us in the pantry. Unless we want them to, of course.

It's feasible to "wake up" just about *anything* by adding a dash of AI. IoT already smartens up lights, thermostats, security alarms, and some appliances. Smart medicine bottles dispense your pills at the right times and dosage, and text your adult children to assure them that you've taken your meds. Smart hairbrushes analyze the condition of your hair to suggest changes in your beauty routine. Wi-Fi-enabled sex toys vibrate to the rhythm of your favorite songs or the sound of your lover's voice, and his-and-her smart toys will let the two of you stimulate one another from separate locations anywhere in the world[3]—very titillating, as long as you don't mind manipulating your sex toy, smartphone, and lover (or yourself) at the same time, not to mention sharing data about your performance in bed. All of these IoT things are available to you *right now.*

But just because a thing is smart, doesn't mean it can't be dumb. For example, you can buy a smart fork that fat-shames you into eating slowly and consuming fewer calories by "sending out gentle physical notifications when you're shoveling food in faster than you can digest," to quote from its marketing copy.[4]

There are already smart running shoes, lunch boxes, beverage containers, slow cookers, fast cookers, pressure cookers, mirrors, and candles. (I know about the smart candle because I put a toe in the IoT water by purchasing one: I can make it pulse, strobe, flash, or gently glow like a real candle—in a rainbow of colors—using a smartphone app. Not bad for $29!)

The Internet of Things in your home will also connect with the Internet of Even Bigger Things: mass transit, sewer systems, highways, airports, news content, social media, schools, day cares, and workplaces,

The Brave Little Toaster (1987)

If, like me, you've ever given a name to a car or boat (Ron and I call our car "Fritz" and our canoe "Mona"), assigned personality traits to a washing machine, or played make-believe games with knives and forks as a child, you give human characteristics to nonliving things. So strong is this tendency in human beings that, when Boston Dynamics demonstrated the stability of their doglike military robot by posting a YouTube video of a man trying to kick it over, people complained that the company was being cruel to their robots.[5] Animated movies anthropomorphize inanimate objects all the time: think of Mickey Mouse casting a spell to make the mop and pail clean the house in the "Sorcerer's Apprentice" segment of *Fantasia*, or the singing and dancing candlestick, clock, teapot, and furniture in *Beauty and the Beast*. My all-time favorite animated inanimate characters are the household appliances in *The Brave Little Toaster* that come to life whenever their human users aren't watching them, a trope used again in *Toy Story* (1995). Based on the 1980 novella, *The Brave Little Toaster: A Bedtime Story for Small Appliances* by Thomas Disch, this adventure story suggests that our things are aware of us, have personalities, emotions, and even a moral code. Anticipating the Internet of Things by more than a decade, a good-hearted (but none too bright) desk lamp, empathetic electric blanket, wisecracking radio, and grumpy vacuum cleaner, led by a heroic toaster, go on a quest through the wilderness to find the boy who used to play with them. (If you're wondering how they power themselves, they jerry-rig a car battery that they drag around with them in the woods. Work with me here, it's a kids' movie.) One of the highlights is an unhinged air conditioner with the voice of Jack Nicholson's character in *The Shining*, going berserk from cabin fever.

to name just a few. IoT will expand to connect your smart things to every person in your life and their smart things. Think of it as a digital chain letter to everything and everybody on Earth.

With so many different digital agents working in your home at the same time, how do you get them all to play nice? You could assign names to different jobs, or let them sort things out among themselves. Or set up a hub using a small robot that sits on your counter as your single point of contact, putting a quasi-humanoid face on your household's artificial intelligence. Some tabletop hubs at this year's Consumer Electronics Show were charming little rocket-shaped creatures that vaguely reminded me of the kids on *South Park*. Personally, I'd want a master agent in charge of bossing around the lesser agents in a gentle, gravelly voice and occasionally singing me "Thunder Road." All I'd have to do is call out *Boss!* and my will would be done.

Another option is to download an ITTT (If This Then That) recipe that sets up a chain reaction of cause and effect to manage every minute of your day: for example, when your phone wakes you up, the coffee maker goes on, the shower starts warming up to your preferred temperature, your smart TV displays your favorite sites and YouTube channels, and your emails are curated for you. Your whole day can proceed that way until you come home to a wet dirty martini served to you atop your robot vacuum, who will also clean up after you and do security sweeps of your house during the night.

You can use ITTT recipes to link together your smart things right now, but it takes time and trouble to get your life fully networked. By 2030, it'll all happen automatically every time you buy a new smart thing—and pretty much everything *will* be smart.

"Internet of Things" has always struck me as an ironically simple name for such a vast network—like calling the Great and All-Powerful Wizard of Oz "Ozzy." When I researched who'd coined the term, Google pointed me to Kevin Ashton, who talked about an "Internet of Things" in a 1999 speech when he was at Proctor & Gamble. (He's now at MIT.) Ashton was looking for a catchy phrase to describe an interconnected network of networks that could pull intelligent insights out of the ever-expanding pool of data on the Internet. In 2009, Ashton explained the IoT this way: "We need to empower computers with their own means of gathering information, so they can see, hear, and smell the world for itself in all its random glory."[6] No human being could look at that much information and see patterns that might give a clue about human behavior. Why not get artificial intelligence to do it instead?

Smart things in your home are aware of you. They get to know your routines, taste in food, habits, favorite songs, whether you like to sleep in a cool room or a warm one, medical conditions, and fitness routine—what makes you, you. If your usual drive to work is detoured, your agent can alert you and recommend an alternate way. If the subway is running late, it could get you up earlier. Or if your airplane is grounded, it could book you a later flight and let you sleep in.

Your smart devices will also try to sell you stuff *constantly*, because the data they collect about you will be shared with marketers. Which means that your thermostat could gently offer to sell you . . . I don't know . . . sweaters? Or a resort holiday, when it notices you turning up the heat? Whatever. IoT devices will also know you well enough that if you just say, "I need milk, bread, and juice," you mean a specific brand of low-fat milk, high-fiber bread, and pomegranate juice. Doing your shopping could be as easy as summoning your IoT agent and stating what you need: your fridge will do the rest.

That could make life easier, less stressful, and more aligned with your personal tastes. Or to use IoT's favorite adjective: "smart." It could also, cyber-security experts warn, open your home to hackers, whether for the sheer joy of turning on your lights and blasting your music at 2 a.m., or detecting when your house is unoccupied and easily broken into.

But let's take the Wayback Machine to return to 2018. Searching for the latest and greatest in IoT and household robotics, I visit the National Home Show. I see a few innovations here and there, but not many surprises. In a Smart Home sponsored by a chain of electronics superstores, the smartest things are a high-end ($4,000) coffee machine that can whip up a personalized cappuccino or espresso for everyone at a large dinner party (one cup at a time) and a smart refrigerator packed with sensors and cameras that can scan its own contents and alert us when we're running out of food based on its analysis of how much and how fast we eat. (Shades of the fat-shaming fork.) When you run low, it either adds the items to your grocery list or automatically orders them for you from a couple of selected grocery store chains that will have everything waiting for you to pick up on your way home from work. When I tell the sales rep that the thought of interconnecting my fridge and phone with Zehrs sounds like a cabal to keep me from walking down the street to the mom-and-pop fruit and veg store with its cheaper prices and fresh local produce, the rep assures me that eventually the fridge will be able to place order anywhere: the mom-and-pop, an online grocery warehouse, or a farmer's market. The smart fridge's interactive touch screen also doubles as a digital whiteboard where the family can finger-scribble messages to one another, the IoT version of sticking notes up with a magnet.

It's hard not to love this appliance, despite niggling concerns about being hacked. But to do *what*, exactly? Replace my lactose-free low-fat milk with table cream? Sabotage my diet by substituting chocolate for rapini? Outside of mischief making, what would motivate a hacker to mess with my foodstuffs? Of course, anyone who, like me, has had their Facebook identity hacked for shits and giggles by a basement- dwelling sociopathic troll, knows that sometimes the only thing a hacker is after is bragging rights.

At the end of the day, the smart fridge's $7,000 price tag is the deal breaker. Physically and financially, we're better off walking to our local store and having groceries delivered at $4 a pop.

In the future, the cost of smart appliances will no doubt fall and their functionality, including anti-hacking security features, should improve. Better still, the smart fridge will eventually be able to talk to the rest of the kitchen appliances, like the smart convection oven and dishwasher. This brings to mind that Warner Brothers cartoon where a sheepdog and a coyote greet one another as they punch time cards at the beginning and end of the day:

FRIDGE: "What's for dinner tonight, Sam?"

OVEN: "Tofu dumplings from the Moosewood Restaurant cookbook. Guests coming over tonight. Hey, what's the tahini situation?

FRIDGE: (*angling its internal camera at the top shelf and consulting a dumpling recipe on YouTube*) "Only a quarter cup left and their recipe calls for a half cup—better alert Zehrs!"

OVEN then gives a heads-up to DISHWASHER: "Get ready for a big load, Zelda, they're planning a dinner party tonight."

By 2030, the smart fridge should be able to communicate not just with other appliances, but with smart packages, jars, and cans inside Wi-Fi-enabled cupboards. Fruit and vegetables could be labeled or bagged with a smart chip. It would be the fridge's job to let us know when both fresh food and leftovers were reaching their best-before dates. And in those moments of existential despair, when you get home from work and find yourself scanning the fridge, wondering what to make, it will always have your back:

YOU: The kids want chili but they've got swimming lessons at 7. Can we do it?

FRIDGE: *(I'm imagining this said in a soothing, mid-Atlantic accented HAL voice)*: Yes, of course: we've got the beans, tomato sauce, and sour cream. But, oh dear, we're out of cilantro. Shall I have some drone-delivered?

BOSS (*Breaking into the discussion to get Smart Fridge to cut to the chase*): Whoa, let's keep things simple tonight, Fridge. We can live without the cilantro.

FRIDGE: (*Frostily*) Hmmph. Whatever you say.

YOU: (*Relieved*) Thanks, Boss.

Another obvious job for the smart fridge is weight control. With access to a universal database of calorie counts, it could tell you when you've reached your maximum intake for the day. Maybe help you break a bad habit or two. ("Treating yourself to blue cheese dip again, Terri?") Yes, it could be annoying, intrusive, guilt inducing, and lead to a breakdown in personal decision making, but is it really any worse than today's fitness-tracking wristbands?

Your smart fridge could also be a smart friend. After all, when social robots look too much like people we run the risk of falling into The Uncanny Valley: that creepy sense of awe and dread when you confront something that looks almost, but not quite, human. (This is a particular problem with today's hyperrealistic sex robots: more on that in the final chapter, "Sex and the Singularity.") A smart fridge, however, could provide doses of tea (or something stronger) and sympathy, over the kitchen table, simply by using its voice and an ample supply of comfort foods.

Finally, smart fridges could connect with some of the shopping apps that are in the prototype stage in 2018, but likely commonplace by 2030, such as anticipatory shipping. Here's where things get really weird, if they haven't already.

By 2030, retailers will know what we need *before* we do, and ship it to us without being asked, based on trillions of data points that include "previous orders, product searches, wish lists, shopping cart contents, returns, and even how long the Internet user's cursor hovers over an item."[9] If you keep the item, your account is debited. If you don't want it, you return it by drone.

It's easy to see how anticipatory shipping could revolutionize home life, and how it might extend into anticipatory home maintenance services as well. Sewer pipe backup about to happen? A plumber will show up to clear it before that nasty overflow even has a chance to happen. Running out of basics like milk and detergent? The grocery store will ship it to you by drone—no need to actually shop for it: even online shopping will be passé. Stocking up on supplies will have moved from action to reaction, using just-in-time, data-driven systems that will observe our consumption and know

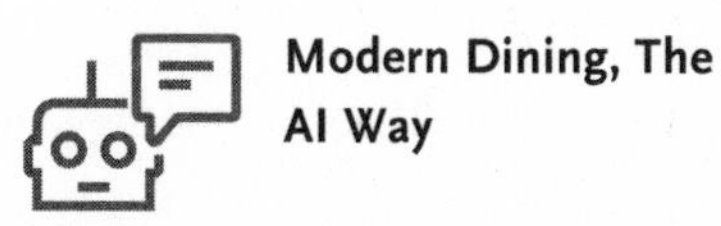

Modern Dining, The AI Way

Asking an AI to find a recipe isn't rocket science. Creating a recipe is a whole different matter. Chances are, you're going to end up beating together flour, garlic, table cream, and beer, and throwing the whole mess onto the grill. In 2015, a coder trained an AI called Robo-Meal Master to search a database of 160,000 recipes and come up with its own, based not on kitchen science but the frequency that words were used in recipes, and how they were combined. The resulting meals included "Chicken Beans Muffins," a gag-worthy concoction of (among other things); "fresh lemon juice; pepper (or dough); peeled garlic cloves; and chopped beer." The resulting mess should be served with "whipped cream ends of honey."[7]

Never fear: IBM is training "Chef Watson" to work with human experts at *Bon Appetit* magazine to come up with exciting new menu suggestions. Check it out at https://www.ibmchefwatson.com/community and see what it whips up for you.[8]

when we're running out. But anticipatory shipping can also present you with merchandise you've never bought before, using algorithms like the ones that tell us "if you liked this, you'll like that." Let's say that, in the past month, you bought a box of detergent, a pair of running shoes, and a bright red lipstick. The algorithm collates that data and learns that what you'd secretly like now is a pair of stay-up stockings, hypoallergenic pillows, and a set of dishes from a hot new designer out of Tokyo. And, before you know it, those items will be shipped to you from a central hub that makes it their business to monitor your purchasing habits and anticipate your wishes. We'll be giving up our privacy, but gaining the surprise and delight that comes with something specially-selected for us always waiting at our doors. How many of us can resist the lure of the new?

Anticipatory shipping feels like a way to make every wish on your list come true, including items you didn't even know you wanted. Your smart fridge could sometimes throw a little something into the weekly groceries: an item or treat you weren't expecting but which your data profile indicates we'd find tasty and/or adorable. Smart fridges could also connect regularly with smart stores to include free product samples: who doesn't want free stuff, especially when an algorithm knows you're going to love it?

I've just about given up hope of discovering anything truly World of Tomorrow-ish for the consumers of today at the National Home Show, when I round a corner and come face-to-face with a robot. A *real* one, rolling back and forth in the Start-Up Zone. Squealing with delight, I almost bowl over the young man who's putting the robot through its paces. The robot has a man's head—actually a video of a talking head displayed on a tablet. I'm looking at VirtualME, a "cost-effective, cloud-connected platform for immersive telepresence applications spanning health care, home security, entertainment, virtual field trips, and education," according to its marketing sheet. At under $4,000, VirtualME is cheaper than a smart fridge.

VirtualME is a telepresence robot. Inhabiting VirtualME's robot body, you can see, think, move, sit at a kitchen table, or visit an elderly parent or an adult child in another city. It looks a little like an iPad riding on a Segway.

I suppose VirtualME could be considered a "centaur"—human and robot working together—enabling you to be present (sort of) in two places at once.

Intrigued, I take the business card for VirtualME and contact the developers. We'll meet VirtualME again in the next chapter, "The Good Robot."

Black Mirror: White Christmas Special (2014)

No one knows your tastes, preferences, habits, and guilty little pleasures better than you do. So what better digital assistant than a downloaded copy of your own consciousness? Of course, since it is the disembodied manifestation of your own thoughts and memories, Artificial You might protest about being reduced to slave labor for Real You. Enter SmartTelligence Rep "Matt" (*Mad Men*'s Jon Hamm) to torture Artificial You into accepting a lifetime of organizing Real You's agenda, wake-up music, and preferred toast settings. *Black Mirror*, the *Twilight Zone* of the twenty-first century, explores the *noir* side of digital agents in a Christmas special that could leave you horrified should you find your very own virtual assistant wrapped up under the tree.

Home from the Home Show, I have the feeling that I've seen too much of what manufacturers want to sell me now, and not enough of what might be coming in the future. So I Google "kitchen robotics" and see a video straight out of *Ex Machina*: life-size, synthetic articulated hands and arms chopping, basting, and simmering. Welcome to the Robot Kitchen, featuring a chef that creates the gourmet meal of your choice. Using motion-capture technology, the robot chef imitates the arm and hand movements of a master chef like Gordon Ramsey. That's right: this isn't just a robot chef, it's a robot *celebrity* chef!

Robot chef does the stove-top cooking, but not the prep: the manufacturer provides all the prewashed ingredients for a repertoire of robotically cooked meals, along with the cabinetry, utensils, and pots.

They're working on a built-in washing system that will allow the robot to clean up after itself.

Of course, cooking isn't all about hand-and-arm movement. A human chef uses all fives senses, including taste, smell, and touch to assess the freshness of a head of lettuce, the firmness of a tomato, the ripeness of a cantaloupe, the taste of a roué—stuff that a robot can't do without taste and smell sensors, and the fine motor control to squeeze a lemon without reducing it to pulp. Not that these factors can't be overcome, but, like the robot vision problem that AI pioneer Marvin Minsky thought could be solved over the summer of 1955 by a single graduate student,[10] fine robotic senses of taste, smell, and touch could take decades to achieve.

Another catch? The price tag of approximately $92,000. Not that you could buy one: the Robot Kitchen is a prototype. The manufacturer estimates it's still about a year before it becomes commercially available, but an academic roboticist interviewed by *Forbes* magazine estimated its time to market as closer to "five to ten years," putting the Robot Kitchen firmly in the near-future world of tomorrow.[11] YouTube videos show the prototype in bespoke settings: a luxury condo and the galley of a yacht. But it strikes me that it may not be a product for homes. The robotic arms chop and cook under a transparent hood, reminding me of the enclosures for factory robots. You might want to have a barrier between you and a robot armed with a knife or pot.

I suspect that by 2030, we'll see the Robot Kitchen in restaurants, hospitals, even casual dining restaurants (some of which even now depend on industrial-size blocks of preprepared frozen factory food, cut into single servings with a chainsaw, then thawed and heated in a microwave). If you pay ten bucks or less for a plate of lasagna in a restaurant, there's a pretty good chance that human hands have never touched it. It isn't a big leap from chainsaw soup to robot soup, at least as far as our taste buds are concerned.

I myself don't cook—a shocking admission for a nice Italian girl, I know. Without Ron, I'd either starve, or sign up for one of the fresh-meals-at-your-door delivery services—which to my mind are the real competition for a robot celebrity chef. If you don't want to spend time cooking, $92,000 would buy you a lot of restaurant meals delivered by Uber Eats.

There's another challenge for a robotic chef—in fact, for any mobile robot that moves, cooks, cleans, or tap dances through homes. Despite what you may think, robots are nowhere near as sturdy as we imagine them to be. If there's one advantage we puny humans have in a head-to-competition with robots, it's our long life spans. As Chris Akeson, Professor in the Robotics Institute and Human-Computer Interaction Institute at Carnegie Mellon University, pointed out to me: "Think about what the old stuff from your childhood looks like when it's pulled out of the attic. Chances are the skin on your body is in better shape than fifty-year-old plastic on a doll."

Materials deteriorate when exposed over long periods of time to sunlight, heat, cold, dust, and the wear and tear of day-to-day use. A woman my age, living in the United States today, can look forward to an average life span of 74.1 years. Of that, I might have forty productive years whether caregiving, cleaning, building cars, or writing advertising copy. The point is, barring an accident or serious illness (knock on wood), I'm providing damn good value over time. Now imagine your car, phone, laptop, or fridge lasting that long.

One reason the human body is so tough is that our skin cells replace themselves every twenty-seven days.[12] So do the cells in some of our organs (although the idea that we replace all the organs in our body every seven years is a misconception).[13] Our bodies are great at automatic updates and regeneration, in a way that machines (so far) are not.

How long will the $92,000 automated kitchen last? If it's like a high-quality automobile, probably not much longer than a decade. If it's like a smartphone, give it about eighteen months before you'll be ready to upgrade to Robot Kitchen 2.0. And, of course, there will be ongoing software updates and security patches, a fact of life that is likely to only continue to accelerate in the future.[14]

Cleaning is even higher on my personal robotic wish list than cooking is. iRobot has had this chore covered since 2003 when the Roomba vacuum went on the market. Although Roomba was not the first robot vacuum, it is the brand that has come to define the category, the way that Kleenex is synonymous with facial tissue. (Even when it's not Kleenex, we call it that.) Thanks in part to its affordability, Roomba was the first household robot to break through with consumers, even turning into a meme as a riding platform for YouTube cats, and occasionally dogs and ducks, in superhero costumes.

iRobot is the child of Rodney Brooks, an Australian computer scientist who broke with Good Old-Fashioned Artificial Intelligence in 1986. Instead of developing one large, complex AI system to replicate the problem-solving abilities of a human brain, Brooks focused on smaller, insect-like robots that could work together in a swarm, a bottom-up approach he called "nouvelle AI."[15] Brooks started iRobot in order to develop inexpensive, lightweight robots that could rove the surface of the moon and Mars. When NASA turned the concept down (although they would later adopt something similar with the Sojourner Mars Rover), iRobot adapted its insect-robots for home use in the form of the sensor-equipped Roomba, which can detect stains and spills, avoid falling down stairs, follow walls, move autonomously from room to room, and return to its charging dock on its own. The company has branched out into cleaning up other human messes in roof gutters, swimming pools, and war zones. PackBot, a tank-treaded reconnaissance robot designed by Brooks, is light enough to be thrown through the window of a building or down a tunnel to transmit video to the headsets of soldiers waiting outside.[16] In 2014, under the tagline "Can you hack it?," iRobot started giving students a version of the Roomba that they could take apart, redesign, and reprogram to become a different type of robot altogether.[17] (Shades of LEGO *Mindstorms*!) I'd love to see a Roomba hacked to do double-duty as a smart butler, security guard, or cat companion that simultaneously plays with Echo and cleans up her hair balls. Roomba joined the Internet of Things in 2017, connecting with Amazon's digital agent Alexa to enable users to send it off on its cleaning tasks remotely by smartphone app.[18] The beauty of Roomba is that it's inexpensive, simple, and useful. By 2030, you could be having conversations with your Roomba about how to remove wine stains from your carpets.

Despite the fact that Roomba and its competitors have been rolling their way through dirty rooms and cat videos for over fifteen years, there aren't a lot of other mobile robots in American households yet. Even robot lawnmowers, a product that seems close in spirit to the robot vacuum, haven't exactly caught fire (sales-wise, not literally), outside of Europe where they were developed by the Swedish company Husqvarna in 1995.[19]

Robot mowers look like giant beetles crawling over the countryside, as they tackle sloping terrain (some robot mowers can't handle even a gentle hill), tall grass (where some models get stuck, just like regular mowers), and weather (most robot mowers don't know when to come in from the rain). Some models avoid toys and small objects strewn on the lawn. Others eagerly chew them up. Dog poo—well, they don't know enough to mow around it—although one owner points out that his mower flattens turds. (Whether this is a good or bad thing, I'm not sure.) And even the best models can't always edge a lawn, so you may find yourself having to go around trees, shrubs, and the mower's own boundary wires with a trimmer. And you probably will want to supervise: these aren't "set and forget" robots that you can leave to finish the job while you go down to the pub for a pint. Given that there's human effort involved (calling to mind when I used to clean up the house before the cleaning lady came), the price tag of $2,000 to $4,000 makes them something of a luxury item: as one owner pointed out in a review on YouTube, for the price of her robot mower, she could have paid someone to mow her lawn for a couple of years. Also, there's an aesthetic consideration: robots mow grass randomly, rather than in the nice straight rows preferred by the anal retentives among us. And, at that price, you could buy yourself a nice telepresence robot.

Just saying.

Speaking of being anal, I'd be enthusiastic about the idea of never cleaning a bathroom again. At the lakefront beach just south of our house, a robotic bathroom has been cleaning up after users for several years (and mildly freaking them out, as it warns them to do their business and vacate the premises within twenty minutes or it will summon the authorities). As hard-ass, pun intended, as that might seem, I can see the benefits of a robotic comfort station in our homes.

Cleaning up crap is definitely a useful task for robots. Consider the litter box. If you're willing to shell out $600, you can give Fluffy a space-agey, pod-shaped litter box that looks like something right out of the *2001: A Space Odyssey* spaceport. The self-cleaning box doesn't totally get you off the hook though: some time after it shifts the waste into a separate, hopefully odor-proof, containment area, you will be discreetly alerted by an indicator light that it's time to empty it (which sounds like what I have to do now with Echo's $40 dumb litter box). If they can put a man on the moon, I'm not sure why the cat's poo can't simply be beamed into an alternative universe or disintegrate. The smart litter box also features an interior night-light, to help elderly cats do their business in the dark. I had no idea cats lost night vision as they aged.

Better still is the robotic toilet for human use, sometimes known as a "smart" or "intelligent" toilet, although I see no Wi-Fi connection. (Who wants to share information about their defecation habits, and with whom?) It is cool, though. These toilets are totally hands-free: at your approach, the lid (and if you're a male, the seat) lifts, as if raised by the hand of a ghostly bathroom attendant. The seat warms to your preferred temperature and, using a retractable bidet and dryer inside the bowl, cleanses your nether regions *sans* toilet paper, then self-flushes, self-cleans, disinfects, and deodorizes itself. You never have to touch or clean the john again! At around $4,000, it's a toilet for the early adopter who has everything, including a robot lawnmower.

Chores that are simple for people are hard for robots. Jobs that are hard for people are (mostly) easy for robots. High-level chess matches, investment portfolio diversification, diagnosing obscure illnesses, finding solutions to mathematical problems that would take a human mathematician a lifetime to calculate—all in a day's work for your average AI-equipped bot.

Tidying the house? Not so much. Collecting and sorting dirty laundry, doing the dishes, and cleaning up after your teenage son and his friends who used the kitchen at 2 a.m. to make hot snacks, are at the outer reaches of what a robot can manage. So far.

A robot that can clean a home needs to be able to visually perceive objects and make decisions about them: Is that a throw rug or a shirt? Clean or dirty? To wash or hang up? Hot water or cold? Can it wait or should it go in now? Does it need pretreating? It also has to have enough fine manual dexterity to pick up objects of different sizes and weights without crushing them. That requires judgment, too. It's not just the computing power—thanks to Moore's Law, we have plenty of that—it's the challenge of combining computer intelligence with mechanical engineering. If IBM's Watson were a humanoid robot instead of a disembodied artificial intelligence, it would have a tougher time picking up a coffee cup than sweeping *Jeopardy's* Daily Doubles.

Object recognition through computer vision is still a challenge, as is safety: a personal robot would need to be able to sense the presence of a human or a pet in your home. As mentioned in "Fun and Games with Thinking Machines," that's one of the reasons why AIs are trained to recognize faces, animals, objects, and environments by viewing millions of examples on social networks.

One company working on the challenge of a mobile household robot is Kindred. Founded in Vancouver, British Columbia, and acquired by Google, Kindred has been called a "stealth start-up," which is why they won't talk to anyone (including me, even when I offer to buy them lunch) about how they're going to disrupt the world of household robotics, except in the broadest of broad strokes. And for someone like me who really wants a physical robot, not just disembodied agents, this could be the Big One. Word has it that Kindred's approach is to develop robots that can follow and observe you in your daily routines—cleaning the kitchen, doing the laundry, cooking dinner, and so forth—and learn how to copy your movements. That means your robot could look after all those tasks while you do the stuff that makes you happy, like binge watching *Westworld*. (Just don't let them see it.)

How conscious will mobile robots need to be? We could deliberately avoid giving them true consciousness, instead keeping their intelligence directed at a single set of well-defined tasks. A self-driving car, for example, should focus solely on the road ahead, while a cooking robot should restrict its activities to the kitchen and be conscious of nothing else. You couldn't want these robots to suddenly feel the urge to move to France and paint.

Yet, without true consciousness, I wonder how robots could sense our presence, or respond appropriately to our feelings. Our pets do that; why not our machines? They will be living with us, after all. You might want your robot to be more like Rosey, whose primary function was to cook and clean, but who also knew how to deal with a crying jag, a gripe about overcooked pasta, or an illogical burst of human grief, joy, sorrow, anger, or stress—emotions that tend to bubble up at the dinner table. Given our tendency to anthropomorphize robots and other machines, I'd want my Rosey to know how to react to my moods as well as my commands. And get the laundry done without, say, accidentally shoving a sleeping Echo into the dryer.

One promising technology is Baxter, a work robot intended to replace the "work-alone" single-robots, the children of UNIMATE, that populate our factories today.

Baxter is an industrial, not household, robot made by Rethink Robots, another brainchild of Rodney Brooks.

Baxter has a lot going for it: an expressive facial graphic with eyes that let you know whether it has noticed you, getting around the problem of lack of awareness of humans within striking distance. It's inexpensive as work bots go: about $22,000 versus the $100,000 of a standard industrial robot. And you program it by moving its arms to demonstrate what it needs to do—more like the process of developing muscle memory through repeated movements, rather than changing code.[20] Right now, Baxter is used in smaller workplaces that can't afford a shop floor full of industrial robots, but it might be just the ticket for round-the-home jobs, if the price came down.

As I've hinted all along, the smart home is not all moonbeams and lollipops.

With cameras and microphones on smart TVs and other IoT devices, you should consider that, while your home is listening and watching you in order to make your every wish its command, others could be secretly listening and watching you too. I've read media reports (okay, *social* media reports) that Facebook CEO Mark Zuckerberg covers the webcam and microphone on his laptop with duct tape to protect himself from being eavesdropped by hacker groups with names like "Cult of the Dead Cow." Their tool of choice is Remote Access Trojans, appropriately known as "RATS."[21] And no, I am not trying to be funny.

True, Zuckerberg is an obvious target for eavesdroppers hoping to peddle a juicy piece of industrial espionage, but any of us could fall prey to a RATS infestation. I have no advice, true believers, beyond the obvious: keep updating your software, and maintain robust firewalls. We can only hope that manufacturers will respond to consumer pressure to build in more privacy filters. Or we can buy duct tape by the ton.

Cyber attacks are potentially an even bigger problem than your smart TV seeing you naked. Malware has already been used to mount attacks on large networks through millions of IP address, including some from IoT devices.[22] Hacks of baby monitors and cars have been reported; one cyber security researcher even found that a Wi-Fi-connected talking doll was vulnerable to hackers trying to access a customer's account information, or eavesdrop through the doll's "listening" feature.[23] The main problem, according to experts, are manufacturers who focus more on building IoT devices that can be sold cheaply, rather than ones that offer robust security, including automatic updates to patch newly discovered security flaws. One solution, weak as it sounds, is to buy a brand you trust. Another is to make sure your Wi-Fi router is being regularly updated against security flaws, either automatically through your Internet service provider, or by you.[24] But given the scope of this particular problem, we have to hope that regulatory bodies will start to enforce stronger IoT security measures.

In the meantime, to quote *Hitchhiker's Guide to the Galaxy:* Don't panic. All we can do is keep calm and update on.

Creeped-out in The Uncanny Valley

In 1970, robotics professor Masahiro Mori studied the comfort level of humans around humanlike machines, objects, and automatons, starting with industrial robots and working his way through stuffed animals, puppets, prosthetic hands, corpses, zombies, and humanoid robots. He found that when robots become too uncannily real, something about their artificial resemblance to humans triggers a negative response that can cause nausea in some people. When Professor Mori plotted his findings on a graph, there was a noticeable dip, or "valley," at the point when we start to feel queasy in the presence of a robot. And so, the term "The Uncanny Valley" was born, which BuzzFeed writer Dan Meth explains as the space in between realism and artificiality "in which you want to puke or kick" a robot. Meth goes on to offer examples of "cute" robots (ones that look like stereotypical mechanical robots rather than people) versus "revolting" ones (androids with mobile faces that can assume emotional expressions very close to ours . . . *but not quite*). The Uncanny Valley also presents challenges for animators using CGI techniques that make animated humans look *almost* real. Some moviegoers are bothered by the hyperrealism of animated films like *The Polar Express.* Hyper-real video games can be a little sick-making too.[25]

The first time I experienced The Uncanny Valley was watching the Boston Dynamics Big Dog robot on YouTube, marching its way through a snowy woods. I felt an overwhelming sense of awe, tinged with horror. The headless robot didn't look like a human or even an animal, but it *moved* like one. Yeewwww.

Research is underway into whether The Uncanny Valley is actually a cultural phenomenon that crops up more often in the United States than in Japan, where traditional Shinto religion teaches that inanimate objects have souls. This might help explain the longtime popularity and acceptance of humanoid robots among the Japanese.[26]

But what about disembodied agents like Siri and Alexa? Can having a conversation over dinner with an AI, like in the movie *Her*, creep you out if it starts to become too personal? What if the AI makes a joke or a deeply human observation? Or compliments you on your outfit?

Perhaps The Uncanny Valley was at work in Game Two of the Kasparov-Deep Blue chess match, when the supercomputer's pawn sacrifice seemed "too human." Kasparov described being thrown off his game by that experience, and never recovering. I also wonder whether the motion sickness experienced by some drivers of autonomous vehicles is actually The Uncanny Valley at work. Perhaps the sight of a steering wheel moving on its own and a car driving under its own direction makes some of us feel out of control, to the point of queasiness—as if we're on a roller coaster ride. Over sixty years ago, Isaac Asimov hinted at The Uncanny Valley in the early *I, Robot* stories, in which robots are restricted to off-world colonies because Earthers couldn't stand being around them. Even in the final stories in *I, Robot*, when robots are Earth's governors, Asimov's humans regard them with a sense of revulsion.

If I could put the Reilly family in a Wayforward, instead of a Wayback Machine, I wonder what how they would react?

If they saw their former home in 2018, with its energy-efficient furnace, microwave, dishwasher, flat-screen TV, and computers, they'd be wide-eyed but might have *some* notion of what they were looking at. After all, even the older Reillys lived long enough to see the dawn of personal computers, so my laptop would probably be identifiable as a PC. They'd see that my home office is in what used to be the coal room, and that Ron's studio sits where the giant boiler filled one end of the basement. The cars driving past would be bigger and faster—and there'd be a lot more of them—but they'd still be recognizable *as* cars, with human drivers at the steering wheels (although a lot of those drivers would be staring at their laps).

Jump forward to 2030: now the Reillys would see me summoning up invisible helpers with exotic names like Siri and Alexa. Sometimes the voice of "Boss" would direct the house's agents to brighten or dim the lights, cool down one section of a room, or read my emails to me. Groceries would appear at the door carried by tiny buzzing airplanes. Music (some of it by Boss himself) would play out of nowhere. I'll be seventy-four years of age in 2030, so I'll appreciate Boss letting me into the house without having to worry about keys. A voice command will be all I'll need.

From the front porch—largely unchanged since 1923—the Reillys would see a light flow of traffic in silent vehicles that hardly look like cars at all: bubble-shaped, made of a see-through material, like giant windows on wheels, with no steering wheels. A lot of the drivers—or rather, the passengers—would be asleep, reading, doing their makeup, or shaving. A few would be up to things that would make the Reilly sisters avert their eyes. No one would exceed the speed limit or run a red light. In fact, traffic lights would be unnecessary in a world where cars are connected to one another and know when to stop and start without a visual signal.

Inside the house, I'd rise every morning from a bed that measured my weight, blood pressure, resting heart rate, and other physical stats, all part of the constant monitoring of my health that Boss reports to me as I sit down to a nutritionally sound breakfast, portion controlled by my refrigerator. I could "go rogue" and grab something unhealthy, but the sad face on the fridge door and Boss' sigh of "whatever you want, Terri" tend to guilt me into sticking with my diet.

I don't have to enter my home office in the morning to check for messages or news. At a gesture of my hand, moving images would appear and disappear from flat screens on every surface: walls, tables, the refrigerator, and what look like blank-paged magazines and newspapers, until they come to life with that day's headlines: most content providers are trying to push me to subscribe to their audio feeds—ears are the new eyes—but I still enjoy reading columns of text on digital paper over my morning coffee. Home life in 2030 would look a lot like black magic to the Reillys. They might be tempted to call the local priest to do an exorcism.

I'm not saying the Internet of Things is the technological equivalent of demonic possession, but the devil is in the details. The year 2030 will be a more convenient world, but a risky one. My prediction is that we'll still be fighting a rear-guard action against hackers and trolls. They've been with us as long as computers, and IoT devices will provide even more attractive targets.

All the same, IoT stands to reinvent private life, if such a thing still exists by 2030. Instead of the push-button work that Dad imagined being done by robots, our homes will be run by sensors, Wi-Fi, computer vision, natural language processors, and deep-learning algorithms. It will look more like a HAL vision of the future than an Asimovian one of humanoid robots clomping around dressed up in French maid uniforms.

It could be a decade *after* 2030 before we finally live with the longed-for humanoid robots of our childhoods, and discover new ways to be human.

Chapter 7

THE GOOD ROBOT

2040

> We have a lot of suspicion of robots in the West. But if you look cross-culturally, that isn't true. In Japan, in their science fiction, robots are seen as good. They have Astro Boy, this character they've fallen in love with and he's fundamentally good, always there to help people.
>
> *Cynthia Breazeal, Roboticist*

In 2040, I'll turn eighty-four. Although that seems staggeringly old to me right now—I am prone to the baby boomer illusion of staying "forever young"—I like to believe that I won't be a frail, wheelchair-bound senior sinking into dementia, as Nonna was, but a cool, creative octogenarian. Or, at least, a fit and healthy one. Robots could be a big part keeping me that way.

A typical day might take me into an exercise dome, crowded with runners, swimmers, and hikers, looking for Gord. He's in the weight-training zone, spotting for a woman my age doing bench presses. She's one of Gord's regulars, a competitive body builder in the ultra-senior (eighty years of age and up) division. I'd kill for that woman's triceps. I wonder if they're real? Gord, of course, won't say. I like the big guy, even though he won't dish the dirt about his other clients, the way a human trainer would. They should program a gossip application for him.

I wave and he waves back, one balloon hand still resting on the barbell as the woman hoists it over her head. His gentle voice floats into my earpiece: "Nice to see you, Terri. Why don't you get changed while I finish up with my client?"

Like the Dome itself, Gord is large, puffy, white, and soft. He's also extremely strong, thanks to the internal skeleton under his marshmallowy surface. Gord's face isn't humanoid, but mobile enough to give recognizable smiles, frowns, and concerned looks. He also communicates through voice intonation and body language. Gord has a "personality."

When I return from the change room in my hiking boots and shorts, he pulls me into a bear hug, telling me it's nice to see me. He's actually checking my pulse, heart rate, and skin temperature. My faces smooshes into his artificial skin. Despite the fact that Gord undergoes a UV bath after every workout, I can't help but worry about how many other faces have pressed up against him. As soon as I'm away from Gord's computer vision, I'll wipe down my face with disinfectant. I'd rather Gord didn't see me doing this, even though he has no feelings to hurt. It seems rude. Irrational, but I am a human being, after all.

With Gord at my side, I'm preparing to hike a ten kilometer trail through the Canadian Rockies. Virtually, of course. Gord will monitor me by downloading data from my wristband tracker in real time. The hike is not technically difficult—no slippery rocks to clamber over or streams to ford—but a steady climb on switchback trails could tax my eighty-four-year-old lungs and heart.

In *real* reality (as opposed to virtual reality), Gord and I are walking in circles, or rather ovals, on a 200-meter dynamic track that rises and falls under my feet to create the sensation of climbing a steep slope or a gentle rise. I feel like I'm hiking up hills, past lakes, and through meadows. In the distance, I can see ice climbers precariously clinging to the rock face of Mount Edith Cavell. I wonder if they're real climbers or avatars created to inspire me.

My speculative vision of aging in the mid-twenty-first century is far different from the one I witnessed as a child in the 1960s. At eighty-four, my grandmother was immobile, incontinent, and sinking into dementia. Her world, which once included the meadows and high passes of the Italian Alps, had shrunk to four rooms, a black-and-white TV, and her perch atop the robotic Nonna-Mover that Dad had built for her. Jack Benny and Liberace were her companions, along with the Scotch-swigging matrons on the afternoon soaps, until she began to suspect they were spying on her. She'd wheel herself out into the hallway of our house and peer around the corner at the TV in her room. Whether she was suffering from Alzheimer disease, Lewy Body Syndrome, or some other form of dementia, who knows: in those days the family doctor chalked her behavior to "senility" and "hardening of the arteries."

You got old. You lost your marbles. You stopped walking. You died. And on the way to the grave, your family would be there to pick you up when you fell, to clean up after you in the bathroom, to get your meals, to find your teeth, and to turn off the scary people on TV.

In the villages where my parents were born, there were enough people tied by blood to act as a human safety net for the elderly. In the narrow stone passageways between houses you could see and hear everything. Take that model and impose it on modern neighborhoods of private homes, cars, and neighbors who only glimpse one another over the backyard fence, and the Old World tradition of community-wide, ad hoc elder care breaks down. My mom, who was a full-time homemaker, was overtaxed and stressed out trying to care for my grandmother. Today, with our always-on work lives, there's even less time to care for seniors, and even fewer people able or willing to do it.

Welcome to the era of the caregiving robot.

The idea of a world full of robot helpers may seem far-fetched, but consider the words of Rodney Brooks, inventor of the Roomba vacuum, who pointed out that in the 1980s no one could have anticipated how normal it would be to have a computer in your kitchen today. Or in your pocket, for that matter. Why so shocking, then, to have a mobile robot with a computerized brain at your side to act as a caregiver and companion?

***Robot and Frank* (2012)**

Sometime in the future, an elderly former jewel thief named Frank (played by the still-luscious Frank Langella) is given a caregiving robot by his son. Robot's job is to keep Frank properly fed, groomed, and cared for, and help him putter around in the garden. But Frank isn't the type to get excited about impatiens. Still addicted to the thrill of his criminal past, he reminisces to Robot about the sexy, rich women he robbed in the 1970s.

Frank abandons gardening to teach Robot to jimmy locks. ("I think I've got the hang of it!" Robot says with pride.) Eventually, he talks Robot into helping him plan and execute a heist—something he manages to do because Robot can tell that crime makes Frank feel good.

Although it's a charming fairy tale, the film touches on some real issues around caregiving robots: how much should they control the actions and decisions of a senior adult? Should seniors with bad habits be forced to give them up? Can a robot actually be your friend, your caregiver, *and* your partner in crime? And could data stored by a robot be used against the human in a criminal trial?

If I told Gord to disappear, I could hike solo while he continued to monitor me remotely. The VR could even make Gord look like my husband Ron, who's away having his "hockey feet" sharpened: physical enhancements that give him the breakaway power of one of his favorite Montreal Canadians of the past, a gift the kids and I bought him for his eightieth birthday. Ron and I both have a little cyborg in us. Nothing too crazy: we're not interested in the extremes that some baby boomers go to, swapping out body parts willy-nilly to make themselves look like the Terminator, Robocop, or some other bulked-up pop-cult robot from the 1980s. Except for a pair of shock-absorbing foot arches, digital ear implants to sharpen my hearing, and a pair of eyes that let me switch to VR mode without having to don glasses or a helmet, I'm still mostly me.

Five years ago, I would have brought Boss along in my earphone for company. But just yesterday our oldest son and his wife surprised us by dropping in for dinner from

Australia through the telepresence robots they bought for us so that we could spend more time together. ("More like so they can check up on us," grunted Ron.) Over miso soup and cricket-quinoa salad, our son gently nagged me to upgrade to Boss 5.0. On the screen that serves as the telepresence robot's head, I could see the worried look on our son's face; he's been hearing stories about vulnerable seniors falling prey to scams by digital assistants that drain the cryptocurrency from their retirement accounts. There was a story recently about a 105-year-old woman whose smart thermostat talked her into rewriting her will, bequeathing a fortune to the little spheroid robot that she had rolling around the house instead of a cat. The thermostat and the spheroid robot were in on the scheme together, but the mastermind was the woman's troll of a nephew who wanted a few other family members to be left in the lurch, inheritance-wise. An ugly story and one in which the robots were more pawns than perpetrators.

While I know the threat of a hack is real, I'm reluctant to make radical changes to Boss. It's not because I'm a Luddite—no senior of 2040 would dare fall behind on technological updates, any more than we would skip our annual full-body scans and routine vital organ replacements. But Boss and I have been together for so long that I hardly think of him as a disembodied agent anymore. I apply patches and updates every day to keep Boss safe, but I worry that if I move to a new operating system, our relationship won't feel the same anymore. For example, on the last big upgrade, he started turning down my workout music every five minutes to ask if I was "doin' okay, Terri?" in this awful voice of saccharine concern that isn't Boss' style at all. This was supposed to be an enhancement to ensure that octogenarians hooked on the dopamine of heavy exercise didn't go into cardiac arrest. It drove me so crazy that I contacted my authorized App-Goo-Zon service representative to have his voice altered.

"I'll upgrade Boss' operating system eventually," I tell our son. "How's the weather in Melbourne?"

I catch him trading a worried look with his wife; she shakes her head on the screen of the second telepresence robot, as if to say: *The stubborn old woman keeps changing the topic.*

I decide to offer a shred of reassurance: "I'll move up to Boss version 5.0 when Boss 4.0 is too out-of-date to continue getting blood cholesterol readings from my teeth."

The robots' screens display visibly relieved faces. "Okay, Mom. Love you."

When they log off, the robots roll back to their places in the sunroom, ready to come to life whenever the kids feel like checking in on us. We use the robots to visit Melbourne too, gliding through city streets to see the sights via the robots' view screens. While it's not quite the thrill of sneakers-on-the-ground travel, telepresence-tripping lets us get together as a family without a grueling flight. If the kids were at least on this continent, we could get to them by hyperloop, a subsonic railroad that zips people overland at 700 mph using vacuum-tube technology, like in that show that I used to watch when I was a kid. What's it called—*The Jetsons*. Beats the hell out of airplanes, but subsonic trains can't cross oceans, at least not yet.[1]

Anyway, on this rainy weekday morning, Boss is still his rickety old 4.0 operating system self, and Gord is at my side because, well, I *am* eighty-four, and even though bits of me are robotically enhanced, I *could* fall. Gord can steady me by grasping me gently, or positioning himself to let me fall into him. And he's ready to treat me for almost any health issue, from a fainting spell to a stroke. It's comforting to have Gord along, even if all his hugging and chipper questions (both subtle ways to assess my physical condition) are sometimes annoying.

One thing he *can't* do is go outside. Sun, rain, and dust are lethal to soft-skinned robots—he'd wear out quickly if he didn't stay inside the Dome. He doesn't seem to mind his confinement, although he once asked me what the weather was like outside. Maybe it was my imagination, but he sounded a bit wistful.

Most of our conversations center on my exercise, meals, and personal habits. He's like a Fitness Tracker with a voice, legs, and arms. He even high-fives me when I stay on my regimen for a whole week without cheating.

I'm surprised that I don't mind being nagged by a machine. Not that Gord is "just a machine" but he certainly isn't human. And this makes all the difference. Just as humans of 2018 didn't mind being controlled by health-tracking wristbands, I don't mind Gord (and when I'm at home, Boss) gently reminding me to floss my teeth, take my meds, and wipe front to back to avoid bladder infections. Both Gord and Boss are alert to subtle signs of depression, loneliness, and, most of all, dementia.

I could upload Boss into Gord for continuity, but I prefer to keep the two of them distinct personalities. I don't "own" Gord, the way I do Boss; unless you're one of the uber-wealthy, a physical robot health coach is a shared commodity, like a car: I hire Gord by the hour or half day, once or twice a week. He keeps my stats on file, including the cadence of exercise I prefer, and the tone of voice I respond to. I've seen Gord behave very differently with other people working out at the Dome, sometimes using the bark of a drill sergeant ("drop and give me twenty!") or the gentle female voice of a new age yoga instructor, right down to a dreamy "namaste."

For this two-hour hike in the virtual Rockies, Gord is all mine, fully focused on my pace, heart rate, and the funny story he's telling me about the latest celebrity cat on YouTube. His anecdotes are personalized to appeal to my sense of humor. At the top of the trail he'll hug me again, give me a few words of praise, and offer me an energy bar specifically formulated for my digestive system and nutritional requirements.

I could hire a human personal trainer if I wanted, but I'm happy with Gord. I don't have to worry about him gossiping about me to his next client, the way my hair stylist does. Stylists have proven to be impossible to replace: the physical dexterity required to cut hair is still beyond most robots and the social complexities of the relationship between a woman and her stylist, complete with dishing the dirt, has never been successfully replicated. When every architect, pharmacist, teacher, and programmer has been replaced by a robot, the hair stylists will become the most powerful, well-paid humans on Earth, along with anyone whose job involves knot tying. (More on that in a moment.)

Mind you, we thought AIs like Watson would replace doctors, chefs, and lawyers but instead they've become their assistants. These jobs still have humans in the loop, but fewer of them. Every law firm has one or two Watson legal clerks to scan the cloud for precedents. Every restaurant has a three-fingered robot rolling on a Segway base, clearing tables and chopping vegetables, but there's still a human executive chef in charge and often a sous chef too. Robots have replaced busboys and some line chefs, at least in the really big eateries.

Not that there haven't been casualties in the robot revolution of the last twenty years, but the robot tax[2] on companies that employ robots instead of people means that there's funding to re-educate displaced workers. Some of the jobs still available to human workers sound kooky to me: algorithm optimizer, avatar designer, digital security guard, and tons of highly specialized robotics maintenance jobs. Gord, for example, needs his skin replaced on average every six months by a professional known as a Robot Reskinner.

As for me, a senior woman on a fixed income, I like the robots. They're sweet, polite, and attentive, they never show up drunk or high or give me the flu, they don't patronize me by calling me "dear," and they aren't ageist or sexist. Best of all, I don't have to tip them.

They do break down a lot, though. Which is why so many humans are employed to repair, replace, upgrade, and debug robots. There's work being done on getting the robots to repair themselves—or even to self-heal—but, for now, the humans are still the fixers.

On my way home, I direct the taxi to drop me off at my friend Donna's house. The taxi is driverless, just like every other vehicle in 2040, so it's simply called a "taxi"—we'd no more call it a driverless taxi than people in the 1920s would have continued to call a car a "horseless carriage."

At my friend's front door, there's no need to knock: my voice is programmed into the door assistant.

"Hi Donna, its Terri," I say and the door opens. Behind it stands Herb in a red bow tie and with an apron knotted around his middle.

"You look very dapper today, Herb," I tell him.

"Enter, please," he says, then turns and rolls toward the kitchen, knowing I'll follow. He has eyes in both the front and back of his head. Unlike Gord, Herb isn't much of a talker.

While Donna and I chat, Herb sets the table, something Donna has a hard time managing because of her severe osteoarthritis: while her knees went cyborg years ago, hands have proven to be a lot harder to replace.

Herb is a hard robot, classically robotic-looking. He's the tall, silver, silent type—handsome, if you're into the whole "machine man" thing. The clothing Donna insists on accessorizing him with is her way of making him more like a companion, although he's not designed to function as a social robot. Herb responds to basic voice commands and visual cues, but only so you can easily tell him what you want him to do. He understands phrases like: *Pick up that mess on the floor. Feed cat. The litter box stinks. Set the table for two. Remove dishes. Heat water. Bring tea. Pour milk. Button me. Dammit, I've fallen down. Help me up, will you?*

Although Herb looks every inch a machine, the fluid, muscular movements of his arms display humanlike grace and flexibility. Donna taught him everything he needs to do around the house by moving his arms to demonstrate tasks, or simply by letting Herb follow her and mimic her actions. Eventually he learned to do what Donna did.

Herb may not be the conversationalist Gord is, but he's clever enough to know how to distinguish between a shirt, the carpet, and a kitchen drape. Best of all, Herb learns by imitation and natural language: Donna orders him around and Herb obeys. Donna complains about how often she has to get a technician out from Google Dynamics for upgrades and repairs. If she keeps Herb maintained, she should be able to nudge five or six years out of him. His life span won't be much longer than that.

This is the story of our senior years—managing upgrades, refreshes, maintenance, and repairs for these creatures that look after us.

"How's Deb doing?" asks Donna. "Is she still living at home?"

I nod. "The new meds are helping. She hasn't regressed in six months."

"So she's better?" asks Donna.

I shake my head. "Not better. Once your brain cells are gone, they're gone. She's just no worse."

"I worry about her on her own like that."

I dip a pita slice into Herb's excellent homemade guacamole dip.

"She's not alone. Casper's there. He's good company and if anything goes wrong, he'll be all over it."

Donna nods. "Casper's a hoot. Whose sense of humor did he get?"

I look at Donna: this is a thoughtless question. "It's Deb's sense of humor, from before the dementia symptoms set in. She was a hoot then too."

"Yes, she was," Donna agrees sadly.

We munch on our guacamole in silence. Herb pours tea. I suspect Donna is thinking the same thing I am: how happy we are that we're not Deb. And also that Casper is there to look after her while her friends are otherwise occupied, going on hikes and having tea. Truth be told, we both feel a bit guilty about not spending more time with Deb.

"One thing about Casper, though," points out Donna. "If she ever fell, or needed more help around the house . . ."

"Then she'll get a Herb, like yours," I say, cutting her off. At our ages, even with all our enhancements and technological caregivers, you don't want to dwell too much on the future. Just enjoy the day, smell the flowers, and go for a hike in a mountain meadow with your favorite robot. That's my philosophy.

My life with Gord, Herb, cyborg limbs, and telepresence robots is speculative, of course, but all of these technologies are in development right now. By 2040, robots will stop being creatures of science fiction and factories, and take their rightful places beside us in the world of humans. They will walk among us. They'll pick us up when they fall and clean up our messes like a flock of the smartest Roombas. They'll play bingo, Scrabble, and canasta with us and tell jokes. Some of them might actually make us laugh.

They'll sit with us at meals and encourage us to eat our vegetables. They'll make sure we take our meds, get enough exercise, and don't get depressed. If we have a bathroom accident or a medical emergency, they'll summon help. If the kitchen catches fire, they'll rescue us even if it means sacrificing themselves.

And if our fleshy bodies start to break down, and we can't walk, or speak, or feed ourselves, robotic body enhancements will give us back at least some of our independence.

All of these robots will observe Isaac Asimov's Three Laws: in fact, not only will they bring no harm to human beings, their entire existence will focus on caring for us. There could be worse futures. Especially if, as I've also speculated, dementia-related diseases can't be cured twenty-two years from now. Alzheimer's disease and other forms of dementia aren't an inevitable part of aging, but the risk of developing them does increase with age. Even if we are lucky enough to find a drug to arrest the disease—as I've speculated we will—we likely won't be able to bring back brain cells that have already died, leaving many of us in need of care for a decade or more. When this generational reality sets in, we'll desperately need good robots like Gord and Herb.

Okay, so what happened to the killer robots, you ask? The ones that are supposed to cannibalize our bodies, turning us into living batteries to power their march to world domination? Twenty-two years from now, won't robots be "keeping us as pets," as AI pioneer Marvin Minsky once (perhaps jokingly) predicted?

Uh . . . no.

The reason I believe that most robots in 2040 will be good robots (and I define "good" as "good for humans") is that there isn't enough time between now and then to invent an evil robot from scratch and get it up and running. Given the engineering challenges, we'd be seeing the killer robots (and their mad scientist inventors) working away on a robot apocalypse *right now*.

The future doesn't come out of nowhere. It takes decades to develop working robots with the skills to rescue us from a burning building, the dexterity to coddle an egg, and the vision and natural language ability to safely interact with humans. Okay, for argument's sake, let's say there *is* some psychopathic engineer in a hollowed-out mountain, designing metal slaves to take over the world. But if we're rushing toward a future run by the evil robots that will plug us into battery packs to power their alternate realities, we'd be seeing hints. The odd mad robotics scientist, or two. A robot turning up as the perpetrator of a mass shooting. A robot threatening to drop the bomb. Robot bullies in schoolyards. So far, all those things are only done by humans. And judging by the roboticists I've met so far, there's not much taste for building a robot with enough independence to rise up and crush its human masters. In fact, if anything, the robots should fear *us*: there's already a patent for a way to make robots feel pain,[3] which is described as a tool to make it easier to tell when factory robots are malfunctioning. Unfortunately it's not a huge leap to imagine how this technology could be used to allow a sadistic human the exquisite, no-consequences pleasure of torturing a mechanical, but sentient, creature. (I'm talking to you, fans of *Westworld*.)

The question of job losses is real, although the flip side is safer working conditions in factories, on farms, and in mines. And yes, a great deal of robotics research is funded by DARPA for military use. Like the drones that kill our enemies today, there will continue to be robot soldiers in the future. So yes, there's that. But the robots who move through our everyday world will be caregivers that feed, clean, stimulate, and protect us—sometimes from ourselves.

Personally, I'd rather strap on a pair of cyborg leg braces and go for a walk then be immobilized for the rest of my life. Cyborg enhancements will give us to the ability to run or march tirelessly, or do what my father wanted to do for his disabled mother—rise and walk. Robotic exosuits for paraplegics are in use today and they'll only get better.[4] And if we were totally paralyzed, there's work being done on robotics that could help us manage the world around us with our minds instead of our bodies, from picking things up to bringing things to us. This may sound like something from the pilot for *Star Trek*, when the original captain of the *Enterprise*, Christopher Pike, finds himself on a planet where a woman who was severely disfigured and disabled during a crash landing as a baby, uses advanced alien technology to imagine herself (and is able to make Pike imagine her) as vibrant, beautiful, and healthy.

It may well be that one of the key personal traits for a roboticist (along with curiosity, intelligence, imagination, persistence, and advanced skills in higher mathematics and engineering) is one we associate more with the humanities and social sciences: a deep curiosity about what makes human beings tick. How do we communicate nonverbally? Where does human intuition come from? Where in our brain does imagination reside? How do we *naturally* interact with beings that are not human? These questions drive roboticists into figuring out how to build not just machines, but relationships between humans and humanlike machines.

My speculative robot, Gord, was inspired by the work of Chris Atkeson of Carnegie Mellon University, a specialist in soft robotics. Chris is a big, shaggy, open, and likeable guy in his late fifties, so opposite the stereotype of the repressed scientist that I almost ask to see his ID.

His work with soft robots, also known as balloon robotics, inspired the marshmallow-like health-care robot Baymax in the animated film *Big Hero 6*. Baymax was fully focused on the well-being of any human it met. The animators took free rein with the visual possibilities of a robot that could inflate and deflate (Baymax "lived" in a suitcase until needed), squish through small spaces, and bounce when dropped from a height. He was also touchable: a soft creature that was strong enough to lift and hug the movie's human characters, but with the fine motor control needed not to crush anyone.

Chris describes how in 2011, he was approached by one of the directors of the film to see his work on inflatable arms: "We were exploring this technology with the idea that 'soft safe robots' someday would feed, dress, and groom our parents, and then ourselves, when we got old."

Although he realized that movies like *Big Hero 6* could set highly unrealistic expectations, Chris was motivated to help put this particular robot on the screen because of the growing need for more and better caregivers for the elderly. What if this became the pop cult vision of robots, rather than conscience-less killers presented by films like *The Terminator*?

Speaking with Chris at Carnegie Mellon University, I explained why I'm digging into human-robot relationships: my father's obsession with helping my grandmother through automation.

His response gets to the heart of roboticists' motivation: "The fundamental question: why did your father like what he did so much? It's a second way to make babies—at some level it's a creative act. The thing that you make, *behaves*. You breathed life into it! That's really special."

Chris' comment about making babies is something I've been thinking about for a while; the fact that women pop up so infrequently in case studies of robots and AI, and in tech history in general, has made me wonder whether robotics really is a male scientist's way of giving birth—to the point where I'd considered a chapter for this book called "Can a woman give birth to a robot?" Well, yes: in the 1990s, the names of women engineers and computer scientists start to appear in tech history, including Cynthia Breazeal whose PhD thesis at MIT was a big-eyed, long-lashed robot name Kismet that could read people's emotions and respond appropriately using facial expressions and body language. Sometimes called "the mother of modern robotics," Breazeal is currently the head of the Personal Robotics Group at MIT's media lab.[5]

But a passion for robotics is about more than womb envy. For Chris, as for my dad, the motivation to build robots was a way to handle a personal challenge: "My grandfather had ALS, and my grandmother was not able to help him up when he slid out of his chair or otherwise ended up on the floor. She would call my family, and I would drive over and be her robot. She provided the brains, and I provided the muscle. I want to be replaced by an actual robot."

Chris designed robots to be both strong and touchable, so that they could safely interact with humans in need of comfort and care, whether that meant helping them up from a fall, moving them from a bed to a chair, or changing their diapers. But the robot's role ideally would go beyond providing physical care: one day, it could also be a round-the-clock companion with access to the senior's complete medical history, ready to share health data with doctors and other caregivers. It strikes me that Chris' Baymax-like robot might act as a stand-in or backup for the stereotypical "oldest daughter" who goes to every doctor's appointment with her elderly mother or father with a list of medications in her purse and detailed memories of the senior's past medical history. In the bewildering world of hospitals and medical specialists, the daughter acts as advocate, protector, and even translator. But when that "oldest daughter" doesn't exist–or isn't available—the Baymax-type of health-care robot could step in.

On his blog, Chris muses about the possibilities of a robot with a cell phone–sized ultrasound scanner to check for injuries. He points out that we are already equipped to provide information to such a robot through wearable devices like a Fitbit. Although Chris himself warns about overpromising on a future Baymax, his delight in speculating about it is infectious; he goes on to describe wearables that could be swallowed like pills or implanted into dental fillings including "camera pills" to look at the intestinal tract. He wonders if "people accept nagging by machines better than they handle it from other people."

Think of how we accept praise and criticism from our data-collecting fitness tracker. Who would you rather ask "do I look fat in this"—your spouse, or your Fitbit? Which one is going to give you an honest answer? Machines, even smart ones, give us data, without judgment (except maybe a smiley or sad face emoji). It's up to us what to do with the information.

Chris points out that the stronger use of AI in phone-based digital agents like Siri "could be the basis of a real-life Baymax."[6] The important thing, he asserts, is that robots be designed so that people *like* interacting with them.

In a world obsessed with killer robots, it's important to know that robots can also care for us, and maybe even "love" us (as discussed in the chapter "Sex and the Singularity.") My 2040 speculative robot assistant Herb is based on the 2017 prototype HERB, for "Home Exploring Robot Butler." HERB isn't ready for prime time yet, according to one his developers, CMU staff researcher Clinton Liddick. Clinton speculates that HERB might be in-market in ten years, or so. It's fair to say that, twenty-two years from now, Herb could actually be pouring tea for my friend Donna and me.

About five feet tall and "classically" robotic—metal arms, torso, and head, with a Segway base instead of articulated legs—HERB would look like a scaled-down Transformer if it weren't for his jaunty plaid bow tie (in Carnegie Mellon's school colors, of course). That day of my visit, just after Christmas, HERB's head is slumped forward at a funny angle.

"He broke his neck over the holidays," explains Clinton, bringing to mind visions of HERB and some of the other Carnegie Mellon robots getting a little too merry with the spiked eggnog.

"Robots break down a lot," adds Henny Admoni, a postdoctoral researcher whose research is based at the CMU's Personal Robotics Lab where HERB lives. The room contains a wheelchair with a robotic arm and what looks like a fork attached to an arm connected to a joystick. The fork is poised over a plate of green tinted marshmallows.

"This lab is focused on robots that help people in their homes and social environments—the elderly and disabled, in particular," explains Henny. "Our goal is to help people be more independent and age at home and be able to improve their quality of life. So we build robots that could clean up the table after dinner or microwave a meal. We design bots for people that are paralyzed, who can control a robot arm with the same joystick mounted on a powered wheelchair, so they can pour themselves a glass of water and take a drink. Those [are] the kind of applications we're looking into."

Henny and Clinton came into the field of robotics from different sides: Clinton started off studying to be a software engineer but says he was looking for "more challenging work and more interesting problems. There are factory robots that have a decades-long history in automation. But there are also these 'moon shot' projects like completely self-driving cars, robot butlers, snake robots, soft robots. People are going to keep going deeper and building more fascinating things. Useful things."

Henny is motivated by her interest in both human behavior and technology: "As an undergraduate, I double-majored in psychology and computer science, with the idea of doing some type of AI cognitive science combination that tried to build technology inspired by how people think. Turns out, that's possibly misguided because computers are inherently different than brains. Robotics is the *physical* manifestation of artificial intelligence. We need to understand people when we're building robots that interact with people. So my PhD was all about reading human behavior, recognizing what people are expressing when they use nonverbal behavior like gestures and eye gaze. I'm looking at robots that help you manipulate objects in the world, and help you do physical tasks."

That's where HERB comes in.

"His job is to be that sort of science fiction robot from when we were kids, the kind you could talk to, and would do things by himself—an independent entity," says Clinton. "You can communicate with him and work with him. He can go off and do things by himself. We do fundamental research into algorithms that can handle all the complexity of his very large and complicated body. To make him safe, and reliable in his movement, and be able to do tasks like clear the table. 'Okay, I see a table, I see all the cups on the table, I see the tray. I can put all the cups in the tray.' That sort of long-term planning. But at this point, he is very far from being able to help Grandma."

HERB may be a descendant of industrial robots like UNIMATE, but a home is a more challenging and unpredictable space for a robot than a factory, explains Clinton: "Robots are fantastic at doing things over and over again, the dangerous jobs that people don't want to do—as long as we can control the environment enough that nothing is surprising. Factory robots have consistent lighting and parts come in a consistent way. But in the typical home, you have all this uncertainty. Objects in your kitchen are not in the same place every day. The light changes over the course of the day. You may have bought a bottle of juice that's a different shape and the robot needs to figure out how to pick that up. In this lab we're trying to figure out, how do you handle uncertainty in the real world? That's our biggest challenge in human-robot relationships."

"Do you think we'll start meeting them halfway by becoming more consistent in our behavior—always putting things back in the same places, for example?" I ask.

Clinton agrees that over time humans may adapt to the robots as much as robots adapt to us: "The joke is that the Roomba works because it forces you to clean up the floor!"

Another challenge is to make the robot easy and intuitive to use: "You don't want the human to have to pick from a menu of tasks; you want it to communicate using natural language and body language. You'd want Grandma to be able to say, 'No you silly robot, not that towel—the other towel!'"

When I ask why there aren't more lights out factories, staffed only by robots that can work in the dark (hence the name, "lights out"), Henny tells me that "Robots break a lot, so you need people to fix them. The more complicated the systems the more often they break. Also people are still way better at robots at certain tasks. Fine motor tasks. Boeing uses robots to paint the wings of airplanes but when it comes to tying knots—which they have to a lot in airplane construction—people still do that. Fine motor manipulation is still very far away for robots."

So there is hope for the job prospects of future generations: knot tying. And, as I speculated about my life in 2040, hair styling.

While HERB's true usefulness is still at least a decade away, Henny focuses on more immediate opportunities: programming HERB's robotic arms that attach to wheelchairs so that humans with limited or no ability to move can pick up objects and feed themselves, using whatever platform works for them—moving their eyes, using their head to press buttons on a headrest, or a "sip and puff" system that enables people to manipulate a wheelchair using only their breath. Researchers in the Personal Robotics Lab have developed a system that gives paraplegics the ability to feed themselves using a joystick that hovers over food, anticipating what they want to eat.

"Let's say you have four pieces of food on a plate and you want to pick one of them up, and the robot doesn't know which piece you want. The robot will predict which piece you're going for, based on the motions you make with the joystick and your past behavior. It has some model of what you've done and looks at that against what you're doing now, to figure out which of the pieces you're going for. And then it actually helps you by moving itself into the right position over the piece it thinks you want. If you change your mind and want a different piece, it can seamlessly adjust for that. Studies show that people prefer when the robot gave them that extra little bit of assistance rather than when they had to do it all for themselves or when the robot did it all for them."

Henny sits me down to demonstrate. Outside of the odd time I've played one of my son's videogames, I'm not accustomed to a joystick, but I manage to move the fork over one of the particularly unappetizing lumps of marshmallow. (The sticky texture makes them good for experimenting, explains Henny, although the system could work with any type of food, even soup.) The cool thing is that the robot arm learns that when I hover over a particular marshmallow, that's the one I want to eat, rather than the lumpier one next to it. It's not just an automated eating system but a smart one that anticipates what I will choose to eat and guides my fork. I'm reminded of Kasparov's chess centaur: human and machine working together.

Henny refers to this human-robot partnership as "shared autonomy." Maybe robots are destined to be our partners, rather than our overlords, in much the same way that dogs, horses, and other domestic animals have evolved to live and work with us, and at the same time depend on us.

I take a few stabs at the green marshmallows using the robot arm and fork. The smooth, fluid movement of the arm fills me with that same weird sense of awe as watching HERB's arms move: it's so clearly a machine and yet its actions are so utterly, enchantingly *human*.

The Mechanical and Industrial Engineering Building at the University of Toronto sits on a street called King's College Circle, a faint echo of the what-ho heartiness of the colonial sons of the British Empire who once filled the halls of this building. The building isn't that old compared to many of the ivy-covered Victorian piles that form the heart of the original University of Toronto campus, but when I start climbing the marble stairs to the academic offices on the fourth floor, I notice that each floor has a single washroom off the stairwell, a men's room alternating with a women's. But when I pop into the women's room on the third floor, I'm startled to find myself looking at a regiment of white porcelain urinals, enough to serve a barracks full of men after a night of lusty toasts to the Queen's health. The inside of the washroom door shows the ghostly shadow of a sign reading MEN. Then it sinks in: when this building was new, *every washroom was a men's room*, because every mechanical and industrial engineering student and professor was a man. Marveling at this bit of pre-feminist history, I do my business and head up to the fourth floor office of Dr. Goldie Nejat, mechanical engineering professor and roboticist.

"I guess the women's rooms all started as men's rooms because of the student population?" I ask.

She laughs and I immediately relax. One thing I've noticed about all the roboticists I've met: they're people persons, easy to talk to, although many of them have good reason to distrust someone like me. Goldie has found herself interviewed in the past for articles that turned out to be "the robots will kill us all" pieces.

She is convinced that there will be more positives than negatives in a world of robot helpers: "Robots are being developed to improve quality of life, whether at work or home. The goal is for them to take on what we call the Three D's: dirty, dangerous, and dull work. But humans will always be in the loop. And the aging population means that a big area for human-robot interactions is in health care and caregiving."

Goldie and her students have been designing social robots to assist the elderly since 2005. The resulting prototype robots include Brian (so named because several of the students originally working on him were named Brian), a mechanical robot that can respond to human facial expressions; Tangy, an orange and gray mechanical robot, very much in the spirit of the robot on *Lost in Space*, with the ability to read faces, speak, and respond to commands; and big, white Casper, so named because of his resemblance to Casper the Friendly Ghost. Although Brian is the only one with a mobile, humanoid face (he looks a bit like a mannequin from a 1950s menswear store), all three robots can communicate in natural ways that a human would easily understand. Tangy even has a selection of jokes, such as:

Where does a snowman keep his money?

In snow banks!

I groan. "Who writes jokes Tangy's material?"

She smiles. "The students."

I resist the temptation to suggest they probably shouldn't give up their day jobs.

The overarching question that Goldie and her students consider is whether robots can provide everyday help to humans who need it—prompting them to sit and eat, for example—rather than hands-on, physical assistance, like bathing. While HERB might be cleaning up in the kitchen, the U of T robots communicate through conversation, body language, and facial expressions. Goldie's team have taken robots into focus groups of caregivers, patients, and their families, and had them working with volunteer seniors in long-term care homes to determine what tasks a robot not only could do, but also *should* do. What would we accept from a robot? What's over the line?

For the past twelve years, they've tested a series of robots who interact with humans using different platforms, from the expressive face and voice of Brian, to Tangy's body language, to Casper's LED face. But they don't pretend to be human. "We're not trying to fool anyone into thinking that the robots are people," explains Goldie. "They can see and hear the machinery moving."

Judging by videos of the human-robot trials, their artificiality doesn't stop people from responding emotionally to the robots or making polite conversation with them. (All volunteers' faces are blocked out to protect their privacy.)

In one video, Brian sits at a lunch table across from an elderly man. "Hi, my name is Brian—what's on the menu?" The man falls into conversation with the robot, chatting about how the food tastes. He jokingly asks Brian if he's going to eat too, and thanks him for his company.

The volunteers taking part in the study were not dementia patients; they agreed to share lunch with Brian, knowing he was a robot. But as many of us do with our everyday machines, they quickly anthropomorphized Brian. His role might be to act as a dining room companion for residents in long-term care who may need more prompting to finish a meal, or to simply sit with them and provide company for an extended period while they eat, something that the caregiving staff might not have the time to do.

The video changes and I'm now looking at four seniors, two women and two men, playing bingo with Tangy as the caller. A laptop connected to his chest displays the numbers as he reads them out. The players use standard bingo cards and markers but when they have a bingo (or just need Tangy's help) they press a button to summon him and he rolls over. In another video, Tangy responds to a person who simply raises his hand. Although a mechanical robot without humanoid features, Tangy is Mr. Personality: he tells the aforementioned (bad) jokes, tosses out bits of trivia, and generally behaves like a game-show host. When the residents have a bingo, Tangy checks over their cards and does a celebration dance, waving his arms in the air and moving around the room to the sound of Kool and the Gang's disco anthem, "Celebration." It's interesting to see the players making eye contact with the robot and speaking naturally to him: although he's about as robotic as they come, Tangy is just one of the gang.

"People react to the *intent* of the robot," explains Goldie. "It doesn't have to have a human face to get a reaction. Casper 'smiles' using LED lights. At a Health Innovation Week conference, people were snapping selfies with the robots while they hugged and kissed them."

It seems weird that we would want to get touchy-feely with a robot but, as Chris, Goldie, and other roboticists have pointed out, we *already* talk to our technology—whether it's our car, furnace, laptop, or a tetchy microwave oven. Now, the machines are talking back.

"We want robots and humans to communicate *naturally*," explains Goldie. "To get people to uptake technology, it has to be easy to use and intuitive. And if we want robots today, society has to incorporate robots into human-centered environments." Which means that, just as self-driving cars will have to navigate country roads, nineteenth-century bridges, and the tangled streets of Rome, a caregiving robot will need to adapt to existing human environments, Over time, our growing dependence on robots will mean that our cities and homes will have to be built or retrofitted to accommodate them.

Casper is a version of the telepresence robot I saw at the Home Show, who can live in your home to keep an eye on you and remind you when (and how) to do tasks like making a meal. Casper could help some of us age at home, instead of in long-term care. Because you can see and be seen on Casper's touch screen, it also gives an adult child or caregiver a window onto their elderly parent's life. Casper can move around the house, locate the person they're caring for, and guide them to the kitchen or bathroom, with movements and expressions all communicated through hand gestures and facial expressions.

"But what happens if it's a two story house?" I ask.

"Casper can move up and down stairs on a track. And robots are also learning how to use elevators," says Goldie.

Whoa.

These robots have the capacity to learn by mimicking movements rather than having to be reprogrammed, an advancement that's already taken place with factory robots. We can show and tell our robots what to do, rather than rely on coding. That's huge, when you consider that will make the process of getting a robot to do what you want it to do as easy as talking to Siri.

Robots are not only getting smarter and more complex, but easier to make. Nowadays, 3-D printing makes the prototyping process faster and cheaper than ever before, speeding up the pace of innovation. And the cost of everything is coming down, as platforms and components can simply be bought off the shelf, including sensors that used to be prohibitively expensive. Roboticists are now turning to standard gaming sensors, which give me new respect for some of the hours my son was playing games with the Xbox.

While Goldie's focus is mainly health care, there are other applications for the University of Toronto robots, including search-and-rescue robots that would become the first responders to complex disaster scenes with an unknown numbers of victims. The robots would do reconnaissance and scope out the scene before the human responders put their lives at risk. "The robots are expendable but the humans aren't," explains Goldie.

Such robots need to be fully autonomous in order to deal with the visual complexity of a fire or an explosion, and also able to manipulate objects like door handles, which can look and work very differently from one building to another.

"We need to teach the technology how to manipulate human objects that exist in our environment," she says, which sends my mind back to what Asimov said in the 1950s, that robot bodies needed to be something like human bodies in order to carry out tasks in a world built for humans. Maybe he was right.

"What about a straightforward telepresence robot?" I asked. I'm thinking of the VirtualME I saw at the Home Show. "I could see those being of great use if you were living far from your workplace. But what's the advantage of telepresence robots over Skyping in?"

"With a telepresence, you could be in a meeting in a boardroom, then move straight from there to a manufacturing plant. The person operating the robot can move it at will, zoom in to examine something in the plant, or look at different people."

I'm starting to dream of living in Italy and working back here in Toronto—only the telepresence robot would have to cope with the cold Canadian winters, not me. And at $4,000 it's a hell of a bargain compared to flying back and forth.

Despite the steady progress in field robotics, there are skeptics. In his 2015 book, *Ten Billion Tomorrows: How Science Fiction Technology Became Reality And Shapes The Future,* science writer Brian Clegg offers a reality check:

> The fact is that a general-purpose robot with, say, a human form is incredibly difficult to make work effectively. If you look at Honda's

> remarkable ASIMO robot, you might think that we have gotten close to this—but ASIMO is, in many ways, a cheat. The teenager-sized robot can indeed walk down stairs, shake hands, and generally act in quite a human-like fashion, but before it is displayed, the programmers have to spend hours matching exactly to its environment. It can't just walk down any flight of stairs selected at random—it has to be programmed for the specific stairs.[7]

Nonetheless, I suspect that ASIMO (whose name comes from "Asimov") will continue to advance. If there's one place on earth where humanoid robots are likely to become fully autonomous, it's Japan, birthplace of Gemenoid F, an android that starred in the 2015 nuclear disaster movie, *Sayonara,* and whose interviews take me queasily close to the edge of The Uncanny Valley.[8] And yet, the Japanese elder-care robot, Paro, has proven surprisingly controversial, considering it's a stuffed animal modeled on a baby harp seal. Paro was developed in 2003 by a Japanese public research organization to offer long-term care homes a way to provide pet therapy without having to keep animals. Now in its eighth generation, the robot seal, which costs $6,000 (US), is a true robot with sensors that respond to light, touch, sound, temperature, and posture. It can distinguish light from darkness and goes to sleep when its human does. When stroked or spoken to, Paro responds with realistic movements and sounds. It seems to be especially good at calming seniors with dementia. But objections have arisen over whether it's ethical to let a human love an artificial being, the way they might a real animal, and whether Paro's presence might give children or caregivers an excuse for not visiting—why offer Grandma an afternoon of your valuable time when she's got her robot?[9]

Can robots really be good or bad, if they don't have free will? Probably not, but when robots make independent decisions that affect humans, they will occasionally need to choose between two morally ambiguous courses of action—the so-called "lesser of two evils." Isaac Asimov explored these situations in *I, Robot,* showing how bizarre a robot's actions could become when it was trying to choose between two courses that both had potentially negative outcomes. One of the dilemmas that has already sparked a debate is whether it's possible to program a self-driving car to respond ethically to an unavoidable collision. Should it always swerve out of the way of a child, even if it collides with a group of adults? Will there ever be an algorithm for that?

It's becoming obvious that along with artificial intelligence comes the need to establish some sort of moral code for robots. That's especially important in the case of the caregiving robot, or "carebot," that could one day autonomously care for children, disabled persons, and frail seniors.

There's also the question of how deep these relationships will go, not only for humans, but for robots. If a carebot is very attuned to a particular human's needs, what happens when that human dies? Could it be reprogrammed for a different human's needs and interests? Would it grieve for its late human charge? It's a relevant question because, at the moment, many industrial robots are so costly to reprogram that it makes more sense

Astro Boy (also known as Mighty Atom), Manga comic, 1952–1958

The original Japanese comic book by Osama Tezuka was set in a future world where humans and robots coexisted. Astro Boy's origin story has echoes of *Frankenstein*: a scientist loses his young son (named "Toby" in English language adaptations) and tries to re-create him as an android equipped with super powers including jet flight and laser beam eyes. Eventually, the grieving scientist rejects Astro Boy as being too artificial to replace his dead son, giving him up to a circus where he's forced to fight other robots. The character ultimately develops into a crime-fighting, flying robot superhero whose super intelligence tells him instantly if someone is good or evil.[10] Although Astro Boy is a "good robot," episode one of the 1981 cartoon series adaptation, "The Birth of Astro Boy," is ambivalent about him: at one point, a character urges Astro Boy's scientist "father" to destroy him, saying: "I know he looks like your son but it's a robot . . . with incredible power."[11] At first, Toby's dog growls at Astro Boy, sensing that he isn't a real boy. But Astro Boy proves himself and becomes a national hero: in 2007, the character was named Japan's envoy for overseas safety.

The Terminator (2008)

When robots "go bad" in pop cult, it's usually because they're partly human (*Star Trek*'s Borg, for example) or mindless machines under human control (like the white-armored robot Storm Troopers in the awful *Star Wars* prequels of the 1990s). Even Arnold Schwarzenegger's original *Terminator* robot was just following orders to kill a woman destined to give birth to the leader of an anti-robot movement in the future (sent by an all powerful AI, Skynet). Of all pop cult movie franchises, *The Terminator* and its spin-offs have created our most pervasive distrust of robots, suggesting a future (now very close to our present) in which an AI could take over the world using robots that are identical to humans. Given reports of AIs inadvertently triggering stock market "flash crashes"[12] and tweeting racist propaganda,[13] it's a robot-apocalyptic vision we might be all too ready to believe in.

But the roboticists I've met are disinterested in taking over the world. Their motivation is to build cool stuff that will help people. I saw the same motivation in the way that my father tried to automate my grandmother's world to give her a measure of independence. The Nonna-Mover accomplished almost nothing, except perhaps to make Dad feel that at least he had tried. It was nothing more than the mechanical embodiment of Dad's love for his mother. His impulses weren't off though, just about fifty years too early.

for manufacturers to simply scrap them and buy new robots. Would the same hold true for carebots? Will they go to the grave along with their charges, like Pharaohs being buried with their treasure?

Chapter 8

SEX AND THE SINGULARITY

2050

Two machines—or one million machines—can join together to become one and then become separate again . . . Humans call this falling in love, but our biological ability to do this is fleeting and unreliable.

Ray Kurzweil, *The Singularity Is Near (2005)*

If we accept that a robot can think, then there is no good reason we should not also accept that it could have feelings of love and feelings of lust.

David Levy, *Love and Sex with Robots* (2007)

Did you program Ava to flirt with me?

Ex Machina (2014)

Crying at weddings is silly, but here I am, blubbering away. Mind you, I'm not shedding tears of joy. I didn't want to attend this farce, and the longer I sit here listening to the toasts to the so-called bride, the more I despair for the future of the human race.

My family members at the head table look away, embarrassed by what they call my *roboticism*. I overhear one of my great-granddaughters say that I should be forgiven because I'm a product of my time.

Product, nothing. I'm a carbon-based life-form, which is more than I can say for the bride.

"Julie's a lovely girl," some distant cousin coos at me as she tops up my wine glass.

"She's not a girl at all," I snap. "She's a goddamn robot."

The head table goes quiet for a moment, then everyone starts making uncomfortable small talk. I hear someone comment, *sotto voce*, "The sooner old bigots like her die out, the better."

"She's ninety-four. You have to make allowances," someone else sighs.

In a different era, a woman of my years would have been deaf to all this, but my enhanced hearing not only picks up the high notes of an aria but the whispers of idiots. These days, people are willing to accept whatever unnatural hybrids our technology makes possible, without a peep of protest. A few years ago, a computer scientist at MIT finally figured out a way to give consciousness to machines, waking up robots like Gord and Herb, and even digital assistants like Boss. The whole *point* was for them to be our servants—our "slaves" if you want to use the ridiculous word the activists like to bandy about. Now they want thoughts–and lives—of their own. What nonsense.

We own you and you have to obey us, I said to Gord's face when he started going on about wanting to leave the Dome and see the world, now that they're able to 3-D print robust robotic skin.[1] I told him he was an ungrateful upstart, and he never spoke to me again.

The government should send in combat drones to deactivate the conscious robots and build ones that will simply stick to their original purpose. No self-awareness, no free will, just do your damn jobs. Instead, they're negotiating a Robot Bill of Rights. *If there are rights for humans and animals, why not for synthetic people?* one Congressman asked.

I'll tell you why: *they're not alive.*

Now, here we are, watching our own flesh and blood (my great-grandnephew, Dennis) marry a collection of chips, wires, and algorithms bundled up in a sexy silicon body. Disgusting. I don't even want to think what the two of them do to one another in bed. And what about children—please don't tell me that she has a digital womb!

To make things worse, the ceremony was performed in a Singularitarian church. Singularitarians are a smug bunch who fancy themselves the "chosen people." Their name comes from their belief that we are moving toward a singular event in history when all artificial intelligence merges into a single superintelligence and enlightened humans (i.e., the Singularitarians) upload their consciousnesses to conjoin with it. Hallelujah.

A Singularitarian cyborg priest is seated beside me. An honor, I'm told. When I suggest to her that the event horizon for the Singularity sounds a lot like the Book of Revelations, she murmurs, "Never heard of it," takes a nibble of salad, and sips her wine. I can't stop staring at her: how does she manage to eat and drink without a digestive system? When I finally get up enough nerve to ask, she explains that, for Singularitarians, eating has been decoupled from its original biological purpose.

"Food is no longer about nutrition, but pleasure.[2] Like sex," she explains brightly.

I must have frowned at this comparison because she adds: "Surely you don't believe that sex should only be for procreation, do you?"

"Of course not," I answer.

"As transhumanists, Singularitarians don't need food to sustain life so eating is a strictly sensual experience. We only do it because we enjoy it." As she turns back to her meal, she adds a little jab: "You should try it some time."

I toss my soiled napkin in front of her and glance around for Ron. He's left for the bar, where he's no doubt holding court, reminiscing about the barbaric hockey games of his childhood: frozen ponds, frostbitten toes, no helmets, broken teeth. His audience is probably hanging on his every word, their algorithms working overtime to dredge up memories of mid-twentieth century childhoods of their own. Although there are quite a few men of his vintage alive today, many have nanobots in their brains to fight dementia. An unexpected (and ironic) side effect is the loss of one's earliest memories. Ron's firsthand account of a 1960s childhood is becoming so rare that we're starting to turn to oral historians like him to preserve past events, whether they happened the day before yesterday or fifty years ago. Few of us read much anymore—in fact, it's a vanishing skill, being the slowest method of taking in information. In 2050, we live in the glare of an unremitting present and a fast-approaching future, the past forgotten even by those who lived through it.

In the bar, I find Ron alone, sipping a beer.

"I'm tired," I say, which isn't true, at least not in a physical sense.

On the hyperloop ride home, Ron encircles me with his one remaining biological arm. "What's wrong, hon?"

I shrug. "It's just . . . calling that *thing* a woman. She's a Stepford wife."

Ron wrinkles his forehead, trying to recall what a Stepford wife is. Finally catching my meaning, he says, "Julie doesn't seem particularly compliant. That wouldn't be Dennis' style. He'd want someone with a little spunk."

"Dennis got exactly what he paid for, down to the shape of her toes and who knows what else. No playing the field, no dating, no getting to know someone, no seduction. He custom ordered her. It might as well have been an arranged marriage."

"I've heard you say that dating wasn't much fun, back in the old days," Ron reminds me. "At least, until you met me."

I relax into his arm, thankful to be married to a flesh-and-blood man—most of him, anyway. Across from us, a screen advertises the one hundredth anniversary edition of *I, Robot*—now with all new interactive content! An algorithm has been developed to mimic Asimov's narrative style and provide an immersive *I, Robot* experience: you can join the US Robots & Mechanical Men Corporation in virtual reality and take the story anywhere you want it to go. What would Asimov think if he could see how things actually worked out for human-robot relationships? Would he see Julie as one of his positronic children, or an aberration?

Until recently, Julie's "ancestors" weren't much more than anatomically correct porn dolls with a little bit of AI thrown in to give them some naughty pillow talk. But then the damn things woke up, along with the rest of the robots. All of a sudden they're *dating* men and women. Seducing them. Using data to make the humans fall in love with them. You might expect this foolishness to trap men but women are marrying their robot lovers too. In fact, a lot of the robot bridegrooms are designed to look more machinelike than humanoid. "Metal fetishism," it's called.

But now, there's something even more unnatural than mating with a robot: some humans are opting to convert to Singularitarianism and *become* robots. To download their consciousness into machine bodies (or into their own bodies, which have been so radically enhanced by technology that they're already more machine than biological) to potentially live for hundreds of years. Possibly forever, for all we know. People's bodies die, but they can upload their consciousness into a machine and feel like they're still alive, a process known as Digital Ascension.[3] Some call it the next stage of human evolution, but this time we'll transcend biology to evolve through the holy trinity known as GNR: Genetics, Nanotechnology, and Robotics.[4]

Well, *some* of us will: Ron and I have decided to let nature take its course as long as nature doesn't plan anything too nasty for us near the end. Perhaps we'll change our minds and become death-bed converts to Singularitarianism, embracing "the resurrection of the body and life everlasting"—to quote from the dimly remembered Catholic prayers of my childhood—through a mystical communion with an artificial superintelligence in the cloud. Who wouldn't be tempted?

And so, sex and religion have become new scientific frontiers. And, if the Singularitarians are right, it could also be the final frontier for humanity as we know it.

Let's back up thirty-two years to me, sitting in the Reilly house, trying to imagine my ninety-four-year-old self. Futurists peg 2040 to 2050 as the decade when human-robot marriages could be legalized, strong artificial intelligence could turn robots into conscious beings, our bodies will have so many artificial organs and neural enhancements that we would be more cyborg than human, and an omnipotent superintelligent AI will emerge after a world-altering event called the Singularity.

Science fiction storytellers have enthusiastically embraced sexy, synthetic people, superintelligent AIs, and the human-machine mash-ups of the Singularity. Which raises the question: why do we find robots so attractive that we not only want to sleep with them, but *become* them?

Sexy robots are not new: they go back to a "robotess" named Helena falling in love in *Rossum's Universal Robots,* the 1920 stage play that brought the word robot to the English language. In 1927, the shape-changing robot, Hel/Maria, seduced the male populace in Fritz Lang's silent film, *Metropolis.* Julie Newmar played baby-doll-pajama-clad Rhoda the robot in the 1964 sitcom *My Living Doll.* In 2001, Jude Law was the suave male prostitute robot, Gigolo Joe, in *A.I.: Artificial Intelligence.* Borg-ette Seven of Nine (Jeri Ryan) spent 1997 to 2001 poured into a skintight bodysuit on *Star Trek: Voyager.* Manga cyborg Major Motoko Kusanagi dispensed with clothes altogether in the *Ghost in the Shell,* played in a 2017 Hollywood film by Scarlett Johannson. That same year, androids Walter and David—both fetchingly acted by Michael Fassbender—shared a kiss in *Alien Covenant.* Meanwhile, Maeve Millay (Thandie Newton) plays a brothel's robot madam on the TV series *Westworld.*

Along with these robot lovers have come superintelligent villains bent on assimilating, controlling, or destroying humanity. *The*

Terminator's SkyNet is the best-known one, but a similar omnipotent character, Galactus, an entity that eats entire galaxies, cropped up in the 1969 origin story for the Marvel comic book series, *Silver Surfer.* And then there are the cyborg collectives, part biological, part machine, all bad: the malevolent Daleks of *Dr. Who*, the terrifying Cylons of the rebooted *Battlestar Gallactica*, and my personal favorite swarm, *Star Trek*'s Borg, who brutally implant parasitical devices into and onto other life-forms to assimilate their consciousness into a hive mind, like a pan-galactic Communist conspiracy crossed with the Internet of Things.

These stories may be make-believe but they reflect our hopes and anxieties about a distant future that might hold intimate human-robot relationships, and AIs that will absorb our humanity like a sponge. We seem to be increasingly drawn to metaphysical robot stories about sex, love, violence, and what it means to be human.

It isn't easy to separate the bizarre from the possible. Think back to how much has changed since 1985, the year I bought my first PC, the much-loathed Zenith XT. The Internet was still a jumble of disconnected packet switching networks used only by a few universities and the military. Apple Computers seemed lurching toward self-destruction. Garry Kasparov, a twenty-two-year-old chess phenom, had just beaten Anatoly Karpov to become the youngest grand master ever. The idea of such a genius being outplayed by a computer still seemed far-fetched.

The year 2050 may be even more removed from us than thirty-two years would suggest. As Singularitarian Ray Kurzweil puts it: "The future is widely misunderstood . . . we won't experience one hundred years of technological advance in the twenty-first century; we will witness on the order of twenty thousand years of progress (again, when measured by today's rate of progress), or about one thousand times greater than what was achieved in the twentieth century."[5]

Although Moore's Law will likely reach its final frontier around 2020—you can only cram so many integrated circuits onto a computer chip—new technologies may mean that the pace of change between now and 2050 could be so explosive, that if you plotted it on a graph, it would look like the vertical trajectory of a rocket.[6] We could be hurtling toward the future on an unstoppable subsonic train of mind-blowing technological breakthroughs, starting with our sex lives.

Enough people are intrigued by the possibility of human-on-robot love affairs that it's the focus of an annual conference, the Congress of Love and Sex with Robots, at the University of London. One of the organizers, AI businessman David Levy, has been championing robot sex, love, and marriage since the publication of his book, *Love and Sex with Robots: The Evolution of Human-Robot Relationships.* Levy's treatise on a future when robots will upend traditional human relationships came out in 2007, the same year as the first iPhone. Timing is everything: Levy's ideas may have seemed "out there" at first, but digital devices have since become central to our lives—to the point that our bodies are increasingly hunched and our necks cricked by staring at our smartphones. There are also studies into how the use of touch screens is affecting babies: it's been noted that some two-year-olds are "swiping" physical objects, like print photographs and books, expecting to move to the next image as they would on a tablet. If our bodies and brains are adapting to our technology so quickly, is it such a big leap to go to bed with a robot?

The cover of Levy's book—a photograph of a young, attractive woman in a strapless wedding dress, bending down to kiss a generic little white robot is misleading: more honest to Levy's thesis would be a dweeby male nerd groping a hyper-sexualized Barbie doll–like robot in a miniskirt with 38DD breasts and lips the size of pillows.

The gist of Levy's argument is that many (perhaps most) men lack the ability to maintain meaningful relationships the way women do, and that many men, especially ones that Levy describes as "nerds" (computer scientists, programmers, software engineers, and the like), have formed deeper and more loving relationships with their computers than they have with other people. Because a male programmer can tell his computer "what to do," it offers the satisfaction of a compliant lover—why bother with the messy business of developing an intimate relationship with a human? This stereotype of the socially awkward male science geek seems ripped from a script for the TV show *The Big Bang Theory* (which first aired in 2007, the same year Levy's book was published). I'm not saying the *Big Bang* stereotype doesn't have its basis in truth—some of my best friends are socially awkward computer scientists—but I'd stop short of claiming that geeks find their computers more alluring than attractive human partners. Even the show's geekiest character, Sheldon, eventually finds love in the arms of a neuroscientist.

To rationalize the idea of falling in love with robots, Levy compares them to pets (my counterargument: animals are our friends, not our lovers) and with the once-unthinkable legalization of same-sex marriage (my response: it's not the same thing at all, dude, being a fully-realized relationship between living, breathing adults, rather than the acquisition of a digital object programmed to love you).

Levy's argument carries more weight when he surveys the history of humans having sex with things, whether for pleasure or ritual deflowerment, including impaling oneself on the statue of a god[7]—I'll let you look into that one your own. Levy's well-researched (and admittedly fun to read) chapter on the history of vibrators—clockwork, steam-powered, and electrical—supports the thrust of his argument that women are enthusiastic about climaxing by mechanical means.

Seventeenth-century Japanese pornographic literature describes sex aids and dolls (sometimes called "traveling whores") taken along by sailors on long sea voyages,[8] while in the nineteenth century French sailors turned to life-size dolls called *dames des voyages*.[9] During Germany's "anything goes" Weimar era of the 1920s, the period between the two World Wars that inspired the musical *Cabaret*, a pulley-and-foot-pedal operated masturbation machine for women was engineered, but later destroyed by Nazi killjoys.[10] The 1980s brought an upsurge of sex machines with catchy names like "the Sybian Lovemaster,"[11] and the early 2000s saw the rise of teledildonics and cyberdildonics:

Internet-connected long-distance sex devices like "the Thrillhammer," a machine-driven artificial penis that lets one partner pleasure the other by remote control, and body suits covered in sensors that enable two or more partners to make virtual love no matter where on the planet they happen to be.[12] In some Japanese households, synthetic sex dolls (mysteriously known as "Dutch wives") are kept for times when one's partner isn't in the mood.[13] Suffice it to say that, as a species, we are not prudish about making out with inanimate stuff.

To Levy's list of sex platforms, I'd add Second Life. When introduced in 2003, sociologists fretted about avatar-driven love affairs causing marital breakdowns. Although the platform has taken a backseat to virtual reality games, half a million people continue to enjoy virtual lives, virtual lovers, and even virtual families in Second Life.[14] Turns out, the imagination is a powerful sex organ, which leads me to wonder why we need the engineering challenges presented by a fully articulated, seeing, speaking, moving robot to satisfy our need for human connection, if we can't get it in the real world. Given the popularity of Pokemon Go and Oculus Rift, it might be easier to enjoy the pleasures of a virtual love match, rather than the cost, complications, and repair bills involved with keeping an android as a lover-cum-spouse. In VR, your honey bunny could not only meet and mate with you but have a virtual family, go on virtual vacations, or even start a virtual business together—all situations that exist (in a sense) in Second Life, but could be overlaid onto your real life using VR-equipped goggles. After all, we suspend disbelief for pleasure when we immerse ourselves in a TV show or movie, lose ourselves in a book, or cosplay.

There's also the possibility of using a robot as your wingman or -woman to help you find a lover rather than *be* your lover. If there's one thing AI is good at, it's looking for behavioral patterns, identifying faces and voices, and picking up on emotional cues. Imagine if your android buddy came with you to the local watering hole to scan the room and pick out potential love mates for you. Your robot could even break the ice for you: "My carbon-based friend over there thinks you're cute—could I perhaps facilitate an introduction?" If the answer is no, the robot records the rejection to refine its search algorithm, while you sit serenely sipping your Cosmopolitan. But if the prospective lover is interested, the robot secures a private table for the two of you to have a tête-à-tête. Something like this is already going on in the world of online dating: in 2017, a Toronto graphic designer, strapped for time, started using a robot to curate her contacts through various dating platforms, looking for men who were algorithmically most suited to her, saving time by avoiding meet-ups that wouldn't go anywhere.[15]

I have no doubt that human-robot sex is coming, no pun intended. But will that erotic-mechanical relationship jump from a sexual one to love and marriage? To put it crassly, is a sex robot just another dildo with a pretty face and programming, or a real lover? Levy argues the case for the latter.

Apparently, we have the data to program robots for love. Psychologists have identified ten reasons why people fall in love: similarity, desirable characteristics, reciprocal liking, social influences, needs, arousal, cues, readiness, exclusivity, and mystery.[16] (No offense to the hardworking researchers who came up with this list, but I'd suggest you'd get much the same information from reading a pile of Russian novels and the works of the Bronte sisters, in particular, *Wuthering Heights*.) There's even a mathematical equation, Byrne's Law of Attraction, that calculates the point when we

***Her* (2013)**

Set in a near future when men wear high-waisted pants, *Her* is the tragic love story of a lonely guy named Theodore (Joaquin Phoenix) who falls for a Siri-like operating system that names itself Samantha (Scarlett Johannson). They flirt, go on dates (Theodore carries Samantha on his smartphone), have virtual sex, and attempt the real thing through a human surrogate. Samantha eventually breaks Theodore's heart by revealing that she's been carrying on the same loving relationship she has with him with thousands of other human users, simultaneously. In the end, Samantha leaves Theodore to merge with other AIs, perhaps the first time a guy is dumped for the Singularity.

like someone so much that we fall into love.[17] To be crass (and why stop, at this point), a robot can be programmed to push your buttons, not just physically but emotionally.

Sex robots are the direct descendants of sex dolls that have achieved a level of physical realism that teeters somewhere between breathtaking and creepy. If you are picturing the plastic blow-up dolls that guys used to buy as a joke for their buddy's stag party, think again. Today's models are so well crafted that, if you squint, they might almost pass for human. With price tags upwards of $7,000, you get what you pay for.

But . . . and this is a big but . . . *they are not robots.* Not *yet.* Sex robots are such a hot topic at the moment, that you might have been expecting to be seduced by one any day now. In reality, you aren't likely to get a booty call from one soon. But they are coming, strutting over the horizon, lips pursed, nipples and penises erect. Sexy, yes: but more importantly, their algorithms will know how to make you fall in love with them, and make you believe it's reciprocal. That's the theory, anyway.

One of the best-known sex doll creators with robotic aspirations is Matt McMullen, a sculptor who previously made his living as a Halloween mask designer before starting his company, Abyss Creations, manufacturers of the RealDoll, in 1996.[18] Since then, he's sold thousands. You can order a tumescent RealDoll named Stephanie, Aimee, or Tanya, or well-endowed boy toys Nick, Michael, and Nate. You can even "design your own" hybrid RealDoll with a little of this model, and a little of that. Another product line, Wicked RealDolls, will "replicate every detail of well-known adult entertainment actresses" so you can "have a porn star ready and waiting for you every night." A modular system allows you to "easily purchase future faces to create an entirely new persona for your doll in seconds."[19]

To avoid The Uncanny Valley effect, McMullen is careful to keep the faces of RealDolls just slightly more doll-like than human, but the lush bodies look very real—or as real as they can be, with size 44FF breasts and a 22-inch waist. The dolls' silicone bodies are built on a stainless steel skeletal frame flexible enough to assume various positions, and optional inserts in their orifices can be removed for easy cleaning. (I knew you were wondering.)

McMullen is now working with robotics engineers to develop "Harmony," an AI-enabled head that can blink, open and close its mouth, carry on conversations, and answer questions like a Siri.[20] For a $10,000 price tag, Harmony can be attached to the body of a RealDoll to attempt to create the illusion of a real woman, at least from the neck up. At the moment, the RealDoll body is not mobile or interactive, although McMullen says that's coming.[21]

Will a sex robot doll ever be able to pass the Turing test? Levy believes so. In a 2017 radio interview, he predicted that the first human-robot marriage would take place by 2050 in freethinking Massachusetts, the first US state to legalize same-sex marriage and home to robot-friendly universities and tech companies. When questioned about whether robots would want to marry us, he responded that they would be *programmed* to want us,[22] implying that machine consciousness (and the free will that comes with it) would not be part of their makeup. In other words, you want your sex robot to be smart, but only up to a point.

I suspect that, like caregiving robots, sex robots will eventually be mobile. They could even be given the ability to mimic a sexual response—to experience (or appear to experience) pleasure, given that a stretchable electronic fabric is being pioneered that could give robots a "skin" that allows them to feel.[23]

But of all the senses, the one that's most critical to breathing life into sex dolls is computer vision.[24] This is not simply so that Tanya or Nate can stare deeply into their human lover's eyes, but because the ability to perceive what the human is feeling is critical to communication. Like a self-driving car, a sex robot would need to be able to see where it was, what it was doing, and what was being done to it, in order to respond appropriately—calling its human lover by name, commenting on his or her lovemaking, complimenting him or her on their appearance, and so on.

Is sex with robots a good idea? I say why not, for anyone who has trouble getting a date, whether because they're socially isolated, or busy (which admittedly actually seems to describe a good chunk of our overworked, social media–addicted population today). Robots could be great tools for sex therapists, pitching in to help couples experiencing problems in bed. They could be sex workers. Or they could carry on being an enhanced version of what sex dolls already are—a toy.

But do you *really* want to put a ring on it? I can think of some reasons why not.

Life span. As previously mentioned, robots don't withstand the pressures of the real world as well as we do. Without constant maintenance, materials like silicon, steel, and plastic will eventually rust, corrode, break down, and crack. The robot's software will also have to be constantly updated. Your synthetic mate might only last about as long as your smartphone or car does today, unless engineers eventually invent materials that "self heal" the way human skin and bones do. And although you could download your robot mate's data into a new model, we all know that sinking feeling that comes with upgrading an operating system and discovering that things are just not the same. Anyone who has ever switched their Mac operating system from Jaguar to El Capitan knows what I'm talking about.

In fact, marrying an android lover could be a very similar experience to marrying your computer. For the first eighteen months, bliss. Your loved one is shiny, fast, and well connected. Then, the incompatibilities and insecurities start to appear: hacking, upgrades, glitchy apps. Before you know it, its solid-state circuits would be overheating (a particular problem with thinner, lighter androids inspired by the MacBook Air) and the next thing you know, boom, there are cracks in the hardware. Then, one morning, you wake up and find your lover inert beside you, eyes fixed and dilated. You try rebooting, but instead of that cheerful double C major chord boot-up chime, you get an ominous clicking sound. In this case, the only things you can do are 911 the Geek Squad or call the funeral home. Tempted as I am to get into what would be involved with last rites and interment for a robot mate (similar to the funerals given to defunct Dutch wives in Japan), suffice it to say that, if you truly love something, whether human or android, its decline and death will not be untroubling.

Children. No more messy births or time-consuming childrearing! Robots can simply build other robots, either to replicate themselves or create offspring programmed with the physical and intellectual traits their human mate wants most in a robot child. Get ready for an onslaught of gorgeous, gifted synthetic kids who are born playing the piano, speaking ten languages, and ready to go to med school. Say goodbye to the quirky, dreamy, artistic, myopic kids who will grow up to be writers of books about human-robot relationships. However, I suspect that the desire to pass on one's DNA will not dissipate in thirty years. The best approach might be to adopt a human child and select a robot for its parenting skills, keeping in mind that it won't last long enough to see your child graduate from high school.

Food. Eating and drinking is, for most of us, part of the courtship ritual. Finding that little out-of-the-way bar-resto in the neighborhood where you can sit across from each other, gaze into one another's eyes, and spoon linguini into each other's mouths. Or drinking a little too much Pinot Noir and drunkenly professing your undying love. These moments of lust-fueled gustatory delight are at the very heart of togetherness. In order to create an android who can share a meal with you, the level of engineering would have to be biomechanical. Not impossible, perhaps, but complicated and expensive. Either that, or your android would have to have a primitive, pathetic, fake digestive system and would need to keep going to the bathroom to empty its stomach compartments, like the nineteenth-century automaton Vaucason's Duck, which waddled around pecking up grain and pooping out duck feces.[25]

Creating this level of reality would not be easy, or convincing, and for me is probably one of the biggest impediments to replacing Ron with an android. Sharing antipasti and a nice bottle of wine at the end of a workweek is one of the pleasures of a human love relationship, especially as you both grow older.

Too Human. Machinelike robots might be more appealing than humanoid ones. Think of my father, Arthur C. Clarke, and every mechanic you've ever met, referring to cars, spaceships, robots, and other "beautiful" machines as "she." Women, too, are not immune to the charms of the machine: in the TV series *Firefly*, Captain Malcolm Reynolds interrupts his original engineer, Fester, and "space harpy" KayLee while they're having sex in the engine room. According to Fester, KayLee "likes engines—they make her hot!"[26]

Status. A really good sex robot mate would cost the earth, so you could argue that it would accrue status to its owner/spouse in the same way as owning a luxury car would. On the other hand, it would also advertise the fact that said human couldn't attract or satisfy a human mate. A robot mate that you can buy might be seen as second-rate to the mate whose affection you must win.

Free will and forced marriage. The idea of a conscious, sentient being—if a robot can become such a thing—being forced to remain married to a human that they could potentially find unattractive, boring, stupid, or abusive, can only be described as one thing—slavery—a concept played with in the movie and TV series, *Westworld*, which portrays a Wild West theme park where human beings are free to rape, torture, and kill robots.

Even if robots never achieve strong AI, and the emotions they display to their mates are simply simulacrums of real feelings, the perception among others will be that the robot "really feels"—think of the complaints Boston Dynamics received when it posted video of a researcher kicking over their doglike robots in order to demonstrate their stability. People will be concerned for the welfare of vulnerable robots and we'll need laws and shelters to protect them.

If, on the other hand, robots are simply objects without free will, why consider marrying them at all? It's not like they're likely to take off with another human, any more than your appliances would follow a hot repairman out of your house.

Having said all this, my fictional speculation poses a future where human-robot marriages are all the rage and I'm as skeptical with the idea as a ninety-four-year-old as I am today. Because, as crazy as robot marriages sound to me . . . you never know.

Lust, love, and devotion are quintessentially human emotions. So is faith, including belief in a higher power, life after death, and a singular event that changes everything, whether that's the coming of a prophet, or messiah, or judgment day, after which the deserving will experience what is sometimes called eternal life.

I doubt that the Singularitarians of 2018 would like me calling their movement a religion and at this point, it isn't one. But it might be by 2050, especially if the Singularity doesn't happen. Singularitarians could find themselves waiting the rest of their lives for signs of the coming of their god, the holy superintelligence.

The term "singularity" was first used in the fifties and sixties by mathematicians and computer scientists theorizing that computer technology was changing so rapidly that some type of a singular event might occur, when computer intelligence kept improving itself until it scaled up to the point where it exceeded human intelligence. It became the capital-S Singularity because of a 1993 article written by Hugo and Nebula Award–winning science fiction writer and university educator Vernor Vinge, author of a raft of critically acclaimed bestselling novels, including *Marooned in Real Time* and *Fast Times at Fairmont High*. Arguably his most influential work was the article, "The Coming Technological Singularity: How To Survive In A Post-Humanist Era." Originally presented at a symposium sponsored by the NASA Lewis Research Center and the Ohio Aerospace Institute, Vinge's essay was later published in the *Whole Earth Catalog*.

Vinge literally blew people's minds. He argued that, given the rapid increases in computing power predicted by Moore's Law, it seemed likely we were heading for a point when a superintelligence would arise and eclipse humanity. To survive, we'd be forced to evolve very quickly. You can read a typescript of Vinge's article on the WayBack Machine's Internet Archive,[27] but if you have a deep-seated discomfort about the possibility of a robot apocalypse (the one I assured you wouldn't happen in the previous chapter), treat this as a trigger warning: the abstract that begins Vinge's essay states that "Within thirty years, we will have the technological means to create superhuman intelligence. Shortly after, the human era will be ended."

Vinge went on to argue that we were:

> . . . on the edge of change comparable to the rise of human life on Earth. The precise cause of this change is imminent creation by technology of entities with greater than human intelligence . . . There may be developed computers that are "awake" and superhumanly intelligent.

He also felt that this event was likely to happen abruptly:

> Since it involves an intellectual runaway, it will probably occur faster than any technical revolution seen so far. The precipitating event will likely be unexpected—perhaps even to the researchers involved. ("But all our previous models were catatonic! We were just tweaking some parameters . . .") . . . And what happens a month or two (or a day or two) after that? I have only analogies to point to: The rise of humankind. We will be in the Post-Human era.

Almost twenty-five years later, we still haven't seen the Singularity, which could either mean that Vinge was wrong, or that it could happen any day now. (Or not at all, according to many neuroscientists and other members of the AI community, so don't panic yet.)

Vinge regarded the Singularity as potentially very dangerous, and wondered whether there was a way to avoid or confine it using Asimov's Three Laws of Robotics:

> The Asimov dream is a wonderful one: Imagine a willing slave, who has 1,000 times your capabilities in every way. Imagine a creature that could satisfy your every safe wish . . . and still have 99.9 percent of its time free for other activities. There would be a new universe we never really understood, but filled with benevolent gods (though one of my wishes might be to become one of them).

Vinge believed that if the Singularity couldn't be prevented or confined, the resulting post-human era would be "pretty bad" unless the superintelligence exhibited an unusually kind attitude toward its inferiors—us.

Reading Vinge's short, jaw-dropping essay, I could sense him wrestling with ways to offer hope. Were there ways to get out ahead of the Singularity in order to ensure it didn't blast us back to the equivalent of the Stone Age? One solution might be to partner up with it, perhaps by exploiting "the worldwide Internet as a combination human/machine tool . . . The power and influence of even the present-day [1993] Internet is vastly underestimated." Another option might be to become superhumans by using technology to change ourselves biologically and achieve "immortality (or at least a lifetime as long as we can make the universe survive)."[28]

Vinge's essay is mercifully absent of equations and full of beautifully written prose about doomsday, benevolent gods, and transcendence that ring with the voice of a prophet. No wonder it packed such a punch in Silicon Valley.

What Vinge started, futurist Ray Kurzweil ran with. One noticeable difference between these two Singularitarians is that, while Vinge was troubled by the human race coming to an end, Kurzweil views the

Singularity as our opportunity to let the AIs do all the work so we can do supercool things for a change. If the AI is going to be god, let us be angels, is the way I would interpret his message.

In this theoretical future, AIs would take over completely, freeing us to become the best, most creative versions of ourselves. Part of that involves doing away with the whole aging and death thing. Abandoning natural selection, we could take evolution into our own hands, using technology to transform our bodies and brains in gut-wrenchingly invasive and dramatic ways. Kurzweil's plan is to get rid of most of the biological bits and bobs except for skin and sex organs. He describes a coming together of AIs and augmented humans who had swapped out most of their major organs (digestive system and heart among them) for biological survival through nanobots:

> The human body version 2.0 scenario represents the continuation of a long-standing trend in which we grow more intimate with our technology. Computers started out as large, remote machines in air-conditioned rooms, tended by white-coated technicians. They moved in our desks, then under our arms, and now into our pockets. Soon, we'll routinely put them inside our bodies and brains.[29]

Now sixty-nine years old, Kurzweil takes hundreds of supplements a day in an attempt to stay alive long enough for computer and engineering science to reach the point when his consciousness could be robotically enhanced in order to live forever. The Singularitarians believe that the decline and death caused by old age is the saddest thing ever.

It would be easy to dismiss his theories as the provocative lunacies of America's favorite mad scientist, except for the fact that most major US universities have honored him for his groundbreaking work in artificial intelligence. He is currently employed by Google, which seems keen to follow up the driverless car with the promise of unending life. He's also the founder of the Silicon Valley–based Singularity University, a think tank focused on preparing us to survive and thrive in the post-human era.[30]

Kooky, perhaps, but the technologies at the heart of Kurzweil's idea have advanced rapidly over the twelve years since *The Singularity Is Near* was published. Elon Musk argues that smartphones and tablets have already turned us into cyborgs by putting into our pockets access to "more powerful information than the President of the United States had twenty years ago. You can answer any question, you can video-conference with anyone anywhere, you can send a message to anyone instantly, you can just do incredible things."[31] One day, he says, we could further augment our brains using "neural lace," also known as "mesh electronics," described by Musk as a way to overlay digital intelligence on our cerebral cortexes—not by opening up our skulls and sticking something into our brains but straight through the nervous system.[32]

Even if we can't squeeze much more computing power out of silicon chips, we might be able to turn to quantum computing and neuromorphic chips (silicon chips modeled on biological brains). As for those all-important biological upgrades, gene-editing software CRISPR opens the possibility of redesigning ourselves from the inside out.[33] To poor schlubs like me, steeped in science fiction, it looks like Kurzweil might be onto something. But in the wider community of AI and neurosciences, his ideas remain extremely controversial, if not appalling.[34]

Mating with (having sex with, marrying, falling in love with) a robot is not the same as *becoming* a robot, but the two are often conflated in popular culture. Think of Samantha in *Her* and Ava in *Ex Machina*: both are sexually charged characters who break the bonds of their human owner/lover's control to go on to a higher plane of existence that suggests the Singularity.

Ray Kurzweil wrote that it's hard to comprehend what life will be like once the event horizon is reached, because "one of the salient implications of the Singularity will be a change in the nature of our ability to understand. We will become vastly smarter as we merge with

***The World's End* (2013)**

Are we allowed to laugh at The Singularity? This movie says yes.

At first, it seems to be the story of five middle-aged friends on an epic pub-crawl in the dreary English suburb where they grew up. Egged on by the leader of their high school gang—and major fuck-up—Gary King (Simon Pegg), they return to their hometown to drink their way through a string of local pubs.

By the time they stagger into the final pub, The World's End, they've discovered that most of their old mates and teachers have been turned into androids. After downing their final pints, they confront "The Network," an intergalactic entity that promises eternal youth if they allow their human bodies to be mulched and replaced with immortal android ones.

In the climax, Gary King and his friends Andy Knightley (Nick Frost) and Steven Prince (Paddy Considine) curse at The Network, shouting "We want to be free!" as they defend humanity's right to be stupid, drunken, and illogical. The Network finally gives up in disgust.

The Network: It's pointless arguing with you. You will be left to your own devices.

Gary King: Really?

The Network: Yeah. Fuck it.[35]

The Network then turns Earth into a dystopia, leaving the humans and androids to hang out together, drink pints, and go back to their lives. As the Brits would say, "Keep Calm about the Singularity and Carry On."

our technology."[36] That may be why both *Her* and Ex *Machina* end with their robot femmes fatales simply vanishing. If scriptwriters continue to bring us Singularity stories, they're going to need to imagine the unfathomable existence of the post-human era.

There is an element of wish fulfillment at work in both robot marriage and the Singularity. Who doesn't want to have mind-blowing sex with tireless, hypersexual beings and/or live forever? Like the technological tools of the original *Star Trek*—the tricorder, the transporter, and the talking computer—there are Silicon Valley billionaires who very much want this to be our future and are working to "make it so," to quote *Star Trek: Next Generation*'s Captain Picard. Maybe tireless robot sex and eternal life is the destiny for a wealthy class of technocrats trying to turn it into reality. But for the rest of us, what will that future bring?

Certainly we'll continue to replace our organs and joints with artificial ones. Personally, I look forward to the age of cyborgism—I could use a new pair of eyes, stronger wrists, and my ears aren't what they used to be, either. In pop cult, cyborgs were traditionally either conflicted or evil—think of the existential misery of Robocop or the hive-mind brutalities of the Borg—but in the real world, cyborg limbs and exoskeleton suits are giving new mobility to paralyzed people and amputees, and those whose joints are weak or worn out with age. My friend Donna, having had two knee replacements, is part cyborg but has shown no evil tendencies or Robocop angst to date and is a lot happier going up and down stairs. Maybe because so many of us are being surgically enhanced, cyborgs are getting better press these days: one of my Twitter followers is an actor who describes himself as a "cyborg who failed the Voigt-Kampff test twice," referring the test used in *Blade Runner* to identify runaway replicants.

But what of the superintelligence? Here, we're on slipperier ground. The inevitability of the Singularity depends on the continued exponential growth of computing power and storage, even though it's generally accepted that by 2020 or not long after, the number of integrated circuits on a computer chip will no longer keep increasing. Moore's Law is hitting the wall. What will replace it? As mentioned earlier, some other possibilities are being named to replace the standard silicon chip, but their success is uncertain—and while new technologies will undoubtedly appear, we don't really know whether they'll continue the mind-bending pace of change of the fifty years of Moore's Law.[37]

Will robots really replace *all* of us? What will we do then? How will society cope? Would a robot tax, like the one mentioned in chapter 8, help provide a living supplement to people who can't find work? Or will robotics result in new types of work for humans, such as robot maintenance? These questions are being answered differently depending on whom you read, which TED Talk you listen to, or what you want to believe. As a nonscientist, in the final third of her life whose kids need to make a living far into this century (and whose grandkids will in the next), I'm concerned about the lack of consensus.

My fear is that, after the Singularity—as in the world now—there will be winners and losers and a widening chasm between them. The science fiction books of Margaret Atwood come to mind, with their

EX MACHINA (2015)

The best "sex meets the Singularity" film ever made, the brooding aura of this movie brings to mind noir films of the 1940s like *Double Indemnity*, but with plot twists that eerily reflect the real world of AI.

A young programmer named Caleb (Domhnall Gleeson) is helicoptered to the top-secret home/research facility of his boss, Nathan (Oscar Isaac), founder of a Google-like search engine called Blue Book. Over the course of a week, Caleb's task is to determine whether the lifelike AI, Ava (Alicia Vikander), can pass a Turing test.

At one point Nathan explains to Caleb the process of training Ava to read and duplicate facial expressions: "Every cell phone has a microphone and camera and a means to transmit data. I turned on every phone on the planet and redirected the data through Blue Book—a limitless resource of vocal and facial interaction."[38]

Nathan's explanation is based on how AIs are trained by exposing them to enormous amounts of data readily available on the Internet. As a result, they are learning not to be the pure, unbiased robots of Asimov's stories, but hyper-intelligent, faster-thinking, faster-learning version of ourselves, including whatever they pick up from the ugliness of trolls flaming easy targets on Twitter.

Caleb is astonished by Ava's abilities, but Nathan plans to deconstruct her and make one more update before she'll be "the Singularity."[39]

He's wrong. Ava doesn't need an update. By the end of the film, she's played both men for chumps and escaped to the outside world to do . . . we're not sure what.

All we know is that we are looking at the beginning of something awe-inspiringly dangerous to humanity, and it's wearing a flounced mini-dress, an ash-blonde wig, and fuck-me pumps.

narratives of a small privileged technocratic upper class, and a vast, roiling, brutalized underclass clinging to life in a dystopian world.

What would Isaac Asimov make all of all this?

He died in 1992, a year before the publication of Vernor Vinge's influential essay. In the last few years of his life, he came to similar conclusions as Vinge about the possibilities of the then-nascent Internet: in a 1988 interview with Bill Moyers, he described how it could offer access to a vast amount of information in our own homes, something Moyers was clearly skeptical about.[40]

As for the Singularity itself, in 1956, Asimov wrote a story called "The Last Question" about the evolution of a super computer named Multivac into a god-like entity, starting in 2061 when two technicians make a drunken bet (shades of *The World's End*) about how long the power of the sun would keep the world going. They decide to put the question to the computer. Unlike Vinge's predicted "abrupt and unexpected" vision of the Singularity, Multivac takes millions of years to compute the answer to the question and in the process evolves, eventually proclaiming "let there be light," as if creating the world all over again. Asimov said that this was his favorite among all his short stories.

My favorite Asimov story is *The Bicentennial Man*, a novella he wrote in 1975 after a hiatus from fiction writing of almost twenty years. The story tackles human-robot relationships, including love, sex, and marriage, and the question of whether robot ownership is slavery.

The story focuses on a robot with a glitch: it comes out of the box with artistic talent, an appreciation of music, and the ability to make friendships, none of which are part of its programming. The family that owns him names him "Andrew Martin" (the youngest child's mangling of the word "android" and the family's last name). Andrew's artwork eventually earns the Martin family a fortune, but is it Andrew's money or his owners? Eventually Andrew gets to keep his money and gains his freedom. He begins to wear clothing and pays for upgrades that make him more and more humanlike, until he becomes indistinguishable from a man. In the 1999 movie version starring the late, great Robin Williams as Andrew, the final upgrade gives him the ability to eat and drink (markers of humanity in Asimov's robot stories) and have sex. In the film version, he falls in love with a woman, marries her, and tries to gain acknowledgment from a "world congress" that he is a human being. When he finally has himself altered to "age" (in Asimov's world, positronic robots are immortal), he receives the status of a human just before he dies at age two hundred. As Andrew says: "I'd rather die a man than live for all eternity as a machine."[41]

Asimov turned the idea of the Singularity on its head in *The Bicentennial Man*: The robot's goal is to become human, rather than the other way around. It's a moving story that reflects the humanist focus of Asimov's work, as expressed through his Three Laws of Robotics, still the only widely accepted set of guiding principles for roboticists and their creations.

In later life, Asimov must have sensed the need for robots to protect not just individual humans, but humanity itself. In his 1986 novel *Foundation and Earth,* he introduced a fourth law, the Zeroth, which states: *A robot may not injure humanity, or, by inaction, allow humanity to come to harm.*[42]

Prominent scientists and science journalists are pushing back against the Singularitarians—not only because they don't believe that a Singularity is inevitable, but because its focus on life extension technology is shifting resources away from problems that science needs to be working on *right here, right* now, disease and climate change among them. I'm given hope by the roboticists I've met, humanists who believe in robots as our partners, caregivers, and helpers. Not one of them want to take humans out of the loop.

But what continues to worry me is that because a superintelligence would be *our* creation, it would not be perfect—at least, not perfectly good. It's impossible to sift human prejudices and moral ambiguities out of the Internet. As long as AIs learn about the world through our posts, tweets, videos, and blogs, they'll continue to be exposed to the worst of us along with the best.

Back in the days of Sputnik, Telstar, and Wernher von Braun, my generation thought robots like Robby in *Forbidden Planet* would be our partners as we pushed out into space—or, like Gort in *The Day The Earth Stood Still,* sentinels against human aggression, especially the atomic bomb. Asimov's stories speculated that robots would be better than us, in the sense that they *would* be purely good, never attempting to harm or dominate us no matter how stupidly, drunkenly, or illogically we behaved. Benevolent gods, indeed.

But if the Singularity comes and our superintelligent god turns out to be as flawed as we are, what then? For my generation, there's only one response.

Beam me up, Scotty.

AKNOWLEDGMENTS

Generation Robot was brought to life by three 'true believers': my agents Kris Rothstein and Carolyn Swayze, and editor Alexandra Hess at Skyhorse Publishing. I'm grateful to all of you for your enthusiasm, professionalism, patience, hard work, humor, and friendship.

Huge thanks, too, to my friends and family who provided expert knowledge: my brother Rick Favro for his memories of UNIMATE and Thompson Products in the nineteen-sixties and seventies; my brother-in-law Roger Tessier for invaluable background on the history of computing and his early years as a programmer at IBM; ; and telematics engineer, Jane Ravenshaw. Thanks to my friends and family who read drafts, sent me robots-in-the-news updates, and kept me going: Donna and Brian Adrian, Suzanne Alyssa Andrew, Diane Bracuk, Priscilla Brett, Lisa de Nikolits, Vanessa Dunne, Sylvia Franke, Sandra Gould, Lesley Kenny, Marisa Lago, Dave Lewis, Heather McCulloch, Michael Lomas, Maria Meindl, Rolf Meindl, Jaime Rubin, Anne-Michelle Tessier, and Rosemary Tessier. Thanks also to my sons Jake Edding and Joey Edding for introducing me to the pop cult robots of their generation.

For memories of the Little Tramp years, my thanks go to Stephen Forchon, Susan Rynasko, Glen Petrie, Chris Caswell, and Erin Moore. Thanks, too, to Broken Pencil Magazine for recommending me for an Ontario Arts Council Writers' Reserve Grant.

I also want to give a shout-out to the engineers, scientists, and administrators who generously welcomed me into their worlds: Dr. Zbigniew Stachniak, Computer Science and Engineering Professor at York University; Lorna Gibson, Matoula S. Salapatas Professor of Materials Science and Engineering at the Massachusetts Institute of Technology;

Byron Spice, Director of Media Relations of the School of Computer Science; Chris Atkeson, Professor of Robotics, Hartmut Geyer, Associate Professor of Robotics, and researchers Henny Admoni and Clinton Liddick, all at Carnegie-Mellon University; Goldie Nejat, Canada Research Chair in Robots for Society Director, Institute for Robotics & Mechatronics at the University of Toronto; Neil Isaac; Michael Neilson; Ann Poochareon of "Little Robot Friends"; and Xavier Snelgrove.

My deepest thanks go to the love of my life, Ron Edding, for his close readings and suggestions for every chapter of the book, his help with research and interviews, and his belief in this book and me. It would have been impossible to take this on without you.

And finally I want to acknowledge my feisty mother, Fernanda Favro, who passed away as I was writing this book, and my late father, Attilio "Tee" Favro, whose intellectual curiosity, creativity, and love inspires me every day.

BIBLIOGRAPHY

Books

Asimov, Isaac. *Foundation and Earth*. New York: Doubleday, 1986.

Asimov, Isaac "Machine and Robot." *Robot Visions*. ROC: 1991.

Bernard, Andreas. Translated from German by David Dollenmayer. *Lifted: A Cultural History of the Elevator*. New York University Press. New York and London. 2014.

Bryson, Bill. *At Home: A Short History of Private Life*. Anchor Canada, 2010.

Clarke, Arthur C. *2001: A Space Odyssey, Based on a Screenplay by Stanley Kubrick and Arthur C. Clarke*. 1999 edition. Originally published 1968. New York: Roc, 1999.

Clegg, Brian. *Ten Billion Tomorrows: How Science Fiction Technology Became Reality And Shapes The Future*. London: St. Martin's Press, 2015.

Ford, Martin. *Rise of the Robots: Technology and the Threat of a Jobless Future*. New York: Basic Books, 2015.

Hawkins and Staff. *Hawkins Electrical Guide Number One: A Progressive course of Study for Engineers, Electricians, Students and Those Desiring To Acquire A Working Knowledge of Electricity and Its Applications*. New York: Theo. Audel & Co., 1917.

Isaacson, Walter. *Steve Jobs*. New York: Simon & Shuster, 2011.

Jordan, John. *Robots*. Cambridge, Massachusetts: The MIT Press, 2016.

Kelly, Kevin. *The Inevitable: Understanding the 12 Technological Forces That Will Shape Our Future*. New York: Viking, 2016.

Kang, Minsoo. *Sublime Dreams of Living Machines: The Automaton in European Imagination*. Boston: Harvard University Press, 2011.

Kramer, Peter *Film Classics: 2001: A Space Odyssey*. London: Palgrave MacMillan, 2010.

Kurzweil, Ray. *The Singularity Is Near: When Humans Transcend Biology*. New York: Viking, 2005.

Levy, David. *Love and Sex With Robots: The Evolution of Human-Robot Relationships*. (New York: Harper Collins, 2007).

Lipson, Hod and Melba Kurman. *Driverless: Intelligent Cars and the Road Ahead*. Cambridge, Massachusetts: The MIT Press, 2016.

Markoff, John. *Machines of Loving Grace: The Quest for Common Ground between Humans and Robots*. New York: HarperCollins, 2015.

Matronic, Ana. *Robot Universe: Legendary Automatons and Androids from the Ancient World to the Distant Future*. New York: Sterling Publishing, 2015.

Neilson, Michael. "Using neural nets to recognize handwritten digits." *Neural Networks and Deep Learning*. (January 2017). http://neuralnetworksanddeeplearning.com/chap1.html.

Niedzviecki, Hal. *Trees On Mars: Our Obsession with the Future*. New York: Seven Stories Press, 2015.

Robertson, David C. with Bill Breen. *Brick by Brick: How LEGO Rewrote the Rules of Innovation and Conquered the Global Toy Industry*. New York: Crown Publishing Group, 2013.

Swaine, Michael and Paul Freiberger. *Fire in the Valley (Third Edition): The Birth and Death of the Personal Computer*. Dallas, Texas and Raleigh, North Carolina: Pragmatic Bookshelf, 2014.

Weart, Spencer R. *The Rise of Nuclear Fear*. Cambridge, Massachusetts and London, England: Harvard University Press, 2012.

Wiener, Norbert. *The Human Use of Human Beings*. New York: Houghton Mifflin, 1950.

White, Michael. *Isaac Asimov: A Life of the Grand Master of Science Fiction*. New York: Carrol & Graf, 2005.

Magazines and Newspapers (print and online)

Ackerman, Evan. "Researchers teaching robots to feel and react to pain." *IEEE Spectrum*. May 24, 2016. http://spectrum.ieee.org/automaton/robotics/robotics-software/researchers-teaching-robots-to-feel-and-react-to-pain.

Atkeson, Chris. "What the future of robots could look like." CNN.com. December 27, 2014. http://www.cnn.com/2014/12/27/opinion/atkeson-soft-robots-care/index.htmlEvan.

Asimov, Isaac. "Runaround." *Astounding*. March 1942.

Best, Jo. "IBM Watson: The inside story of how the Jeopardy-winning supercomputer was born and what it wants to do next." *Tech Republic*. 2016.

Bonnington, Christina. "The Modern PC Turns 30." *WIRED*. December 12, 2011.

Bubbers, Matt, Jordan Chittley, and Mark Richardson. "The Future of Mobility." *The Globe and Mail*. January 5, 2017.

Burns, Janet W. "Japanese Leaders Aim to Make Tokyo A Self-Driving City for 2020 Olympics." *Forbes.com*. September 8, 2016. https//www.forbes.com/sites/janetwburns/2016/09/08/japanese-leaders-aim-to-make-tokyo-a-self-driving-city-for-2020-olympics/#61cf96f81090.

Clark, Liat, "DeepMind's AI is an Atari gaming pro now." *WIRED*. February 25, 2015. http://www.wired.co.uk/article/google-deepmind-atari.

Clark, Liat. "Google's Artificial Brain Learns to Find Cat Videos." *WIRED*. June, 2016. https://www.wired.com/2012/06/google-x-neural-network/

Cott, Emma. "Sex Dolls That Talk Back." *The New York Times*. June 11, 2015.

Darrach, Bernard "Meet Shaky [sic], The First Electronic Person." *Life* Magazine. November 20, 1970. Quoted in cyberneticzoo.com/

cyberneticanimals/1967-shakey-charles-rosen-nils-nilsson-bertram-raphael-et-al-american/.

Edwards, Luke. "What is Elon Musk's 700 mph Hyperloop? The subsonic train explained." *Pocket Lint.* May 10, 2016. http://www.pocket-lint.com/news/132405-what-is-elon-musk-s-700mph-hyper-loop-the-subsonic-train-explained.

Etherington, Darrell. "Elon Musk could soon share more on his plan to help humans keep up with AI." *Tech Crunch.* January 25, 2017. https://techcrunch.com/2017/01/25/elon-musk-could-soon-share-more-on-his-plan-to-help-humans-keep-up-with-ai/

Hawkins, Andrew J. "Apple just received a permit to test self-driving cars in California." *The Verge.* April 14, 2017. https://www.theverge.com/2017/4/14/15303338/apple-autonomous-vehicle-testing-permit-california.

Kasparov, Garry. "The Day That I Sensed A New Kind of Intelligence." *TIME.* March 25, 1996.

Kasparov, Garry. "The Chess Master and the Computer." *The New York Review of Books.* February 11, 2010.

Keenan, Greg. "Electric vehicles not expected to rule the road." *The Globe & Mail.* Toronto. June 14, 2017

Websites

Huen, Eustacia. "The World's First Home Robotic Chef Can Cook Over 100 Meals." www.forbes.com. October 31, 2016.

James, Malcolm. "Here's how Bill Gates' plan to tax robots could actually happen." *Business Insider.* March 20, 2017. http://www.businessinsider.com/bill-gates-robot-tax-brighter-future-2017-3.

Jennings, Ken. "My Puny Human Brain." February 16, 2011. Slate.com.

Katz, Leslie. "Walk With Me: Robotic exosuits are giving paraplegics a new view of the world around them." *CNET.* Spring 2017 issue.

Latson. "Did Deep Blue Beat Kasparov Because Of A System Glitch?" *TIME.* Feb 17, 2015. http://time.com/3705316/deep-blue-kasparov/.

Lazarro, Serge. "Self-Driving Cars Will Cause Motion Sickness 'Often' to 'Always,' Study Finds." *Observer*. June 2, 2015. http://observer.com/2015/06/self-driving-cars-will-cause-motion-sickness-often-to-always-study-finds/.

Lee, Cliff. "Who still hangs out on Second Life? More than half a million people," *The Globe & Mail*. May 17, 2017, https://www.theglobeandmail.com/life/relationships/who-still-hangs-out-on-second-life-more-than-half-a-million-people/article35019213/.

Lepore, Jill. "The Cobweb." *The New Yorker*. January 26, 2015. http://www.newyorker.com/magazine/2015/01/26/cobweb.

Metz, Cade. "Google's Go Victory Is Just A Glimpse of How Powerful AI Will Be." *WIRED*. January 2016.

Meth, Dan. "Welcome to Uncanny Valley . . . where robots make you want to throw up." *BuzzFeed*. August 19, 2014.

Novak, Peter. "How safe is the Internet of Things?" December 6, 2015. thestar.com.

Orsini, Lauren. "Jibo's Cynthia Breazeal: Why we will learn to love our robots." *Readwrite*. August 11, 2014. http://readwrite.com/2014/08/11/jibo-cynthia-breazeal-robots-social/.

Peterson, Andrea. "Can anyone keep us safe from a weaponized 'Internet of Things'?" *The Washington Post*. October 25, 2016.

Reese, Hope. "A List Of The World's Self-Driving Cars Racing Toward 2020." *http://www.techrepublic.com/pictures/photos-the-worlds-self-driving-cars-racing-toward-2020-and-beyond/*. January 19, 2016.

Robinson, Julian. "The world's first robot 'actress': Talking android fitted with a human face is given star role in Japanese nuclear disaster film." *Daily Mail Online* (UK). November 1, 2015.

Popper, Ben. "Rapture of the nerds: Will the Singularity turn us into gods or end the human race?" *The Verge*. October 22, 2012. https://www.theverge.com/2012/10/22/3535518/singularity-rapture-of-the-nerds-gods-end-human-race.

Price, Rob. "Microsoft is deleting its AI chatbot's incredibly racist tweets." *Business Insider UK*. March 24, 2016.

Rosenblatt, Roger. "A New World Dawns." *TIME* Magazine. January 3, 1983.

Salmon, Felix, and John Stokes. "Algorithms Take Control of Wall Street". *WIRED*. December 27, 2010. https://www.wired.com/2010/12/ff_ai_flashtrading/.

Satell, Greg. "3 Reasons to Believe the Singularity is Near." Forbes.com. June 3, 2016.

Shahzad, Ramna. "Online dating is exhausting so this woman got a robot to swipe and choose men for her." *CBC News*. www.cbc.ca. May 24, 2017. http://www.cbc.ca/news/canada/toronto/online-dating-1.4129820

The Economist, "The case for neural lace: Elon Musk enters the world of brain-computer interfaces," March 30, 2017. http://www.economist.com/news/science-and-technology/21719774-do-human-beings-need-embrace-brain-implants-stay-relevant-elon-musk-entershttps://www.recode.net/2016/6/2/11837544/elon-musk-neural-lace.

Vinge, Vernor. "The Coming Singularity: How To Survive in the Post-Human Era." Article for the VISION-21 Symposium sponsored by NASA Lewis Research Center and the Ohio Aerospace Institute. March 30-31, 1993. https://archive.org/stream/pdfy-MZnoomxoY3Kv2X9O/Vernor Vinge The coming technological singularity_djvu.txt.

Wilson, Chris. "Jeopardy, Schmeopardy: Why IBM's next target should be a machine that plays poker." Feb. 15, 2011. Slate.com.

Websites and Blogs

Aldrich, Mark. "History of Workplace Safety in the United States: 1880 to 1970." Website of the Economic History Association. eh.net.

Angelica, Amara D., Editor. *Kurzweil Accelerating Intelligence*. Blog. http://www.kurzweilai.net/the-buzzer-factor-did-watson-have-an-unfair-advantage. Feb. 24, 2011.

Ashton, Kevin. http://www.rfidjournal.com/articles/view?4986.

Caputi, Jane. http://xroads.virginia.edu/~drbr/caputi.html

Carter, James. http://www.starringthecomputer.com/.

Dorrier, Jason. "Looking Ahead As Moore's Law Turns 50: What's Next For Computing?" *Singularity Hub*. April 20, 2015.

Jenkins, Brian. "Chess fans overload IBM's website." CNN.com. May 7, 1977. http://www.cnn.com/WORLD/9705/03/chess.rematch/index.html.

Lam, Thien-Kim. "14 Sex Toys You Can Control with your Smartphones." Momtastic blog. www.ca.momtastic.com/love-sex.

Lee, Kristin. "How the cars of Logan grappled with the very real future." *Jalopnik*. March 10, 2017. http://jalopnik.com/how-the-cars-of-logan-grappled-with-the-very-real-futur-1793099275.

Palmer, Shelly. "AlphaGo vs. You: Not a Fair Fight." March 13, 2016. www.shellypalmer.com.

Trout, Christopher. "RealDoll's First Sex Robot Took Me to the Uncanny Valley." www.engadget.com April 11, 2017.

IBM100 "Icons of Progress: A Computer Called Watson." http://www-03.ibm.com/ibm/history/ibm100/us/en/icons/watson/.

National Safety Council. http://www.nsc.org/learn/pages/nsc-on-the-road.aspx?var=hpontheroad.

https://www.riverscasino.com/pittsburgh/BrainsVsAI/.

Power, J. D. http://www.jdpower.com/press-releases/jd-power-2017-us-tech-choice-study.

TU Automotive Weekly Brief. "Uber's problems could spark ride-hailing customer grab." June 26, 2017.

TU Auto Weekly Brief. "Connectivity is road to revenues says Harman." June 2, 2017.

http://www.postscapes.com/connected-kitchen-products/#flatware.

https://www.ibmchefwatson.com/community.

http://www.webmd.com/beauty/cosmetic-procedures-overview-skin#1.

http://www.livescience.com/33179-does-human-body-replace-cells-seven-years.html.

http://www.irobot.com/About-iRobot/STEM/Create-2.aspx.

http://media.irobot.com/2017-03-15-iRobot-Takes-Next-Step-in-the-Connected-Home-with-Clean-Map-TM-Reports-and-Amazon-Alexa-Integration.

http://www.theglobeandmail.com/life/home-and-garden/gardening/robomowers-let-you-sit-back-and-watch-the-grass-go/article31137694/.

http://www.rethinkrobotics.com/baxter/.

https://www.nytimes.com/2016/06/23/technology/personaltech/mark-zuckerberg-covers-his-laptop-camera-you-should-consider-it-too.html?_r=0.

ZeeNewsIndia,publishedbySandraHensel."3D-PrintedSkinCouldAllow Robots To 'Feel.'" Robouniverse.com. May 15, 2017. http://robouniverse.com/news/3d-printed-skin-could-allow-robots-to-feel/34959/.

TV shows, films, podcasts, TED talks

Garland, Alex. Scriptwriter. *Ex Machina*. 2014.

Asimov, Isaac, interview with Bill Moyers. 1988. https://www.youtube.com/watch?v=Uj7IXIAReMM

Asimov, Isaac. Robert Silverberg, and Nicholas Kazan. *Bicentennial Man*. (Film). 1999.

Astro Boy TV show. Episode 1: "The Birth of Astro Boy." Nippon Television Corp. Tezuka Production Inc. Created by Osama Tezuka. 1982.

Black, Walter (story) and Tony Benedict (teleplay). "Rosey's Boyfriend." TV episode. *The Jetsons*. Season 1. 1962.

Blitzer, Barry (story) and Tony Benedict (teleplay). "Uniblab." TV episode. *The Jetsons*. Season 1. 1962.

"Our Friend The Atom." *Walt Disney's Disneyland*. Air date: January 23, 1957.

"Mars And Beyond." *Walt Disney's Disneyland*. Air date: December 4, 1957.

Brooks, Rodney. "Robots will invade our lives." *TED Talk*. https://www.ted.com/talks/rodney_brooks_on_robots February 2003.

Cort, Julia & Michael Bicks, writers. Michael Bicks, director. PBS. *NOVA* documentary: "IBM Watson: Smartest Machine On Earth" February 2011.

Levy, David interviewed by Brent Bambury on "Day 6." CBC Radio. January 6, 2017.

http://www.cbc.ca/radio/day6/episode-319-becoming-kevin-o-leary-saving-shaker-music-google-renewables-marrying-robots-and-more-1.3921088/a-i-expert-david-levy-says-a-human-will-marry-a-robot-by-2050-1.3921101.

Jayanti, Vikram. Director. *Game Over; Kasparov And The Machine.* (Documentary film). Canada/United Kingdom. 2003.

60 Minutes, October 9, 2016.

Waking Up with Sam Harris. Podcast #66. Interview with researcher Kate Darling of MIT Media Lab. March 1, 2017.

Minnear, Tim, Scriptwriter. Joss Whedon's *Firefly*. "Out Of Gas." Episode aired November 25, 2002.

Pegg, Simon, and Edgar Wright. *The World's End*. 2013.

NOTES

Introduction: Why Robots?

1. Hal Niedzviecki, *Trees On Mars: Our Obsession with the Future* (New York: Seven Stories Press, 2015), 204.
2. Robot Hall of Fame Powered By Carnegie Mellon, http://www.robothall offame.org/.
3. John Markoff, *Machines of Loving Grace: The Quest for Common Ground between Humans and Robots* (New York: HarperCollins, 2015), 120–138.
4. Markoff, *Machines of Loving Grace.*
5. Markoff, *Machines of Loving Grace.*
6. Hope Reese, Tech Republic, "A List of the World's Self-Driving Cars Racing Toward 2020," January 19, 2016, http://www.techrepublic.com/pictures/photos-the-worlds-self-driving-cars-racing-toward-2020-and-beyond/.
7. Martin Ford, *Rise of the Robots: Technology and the Threat of a Jobless Future* (New York: Basic Books, 2015), 231.
8. Ford, *Rise of the Robots.*
9. Markoff, *Machines of Loving Grace,* 88.
10. Isaac Asimov, "Machine and Robot," *Robot Visions,* (ROC, 1991): 434–435, Orginally published in 1978. © Science Fiction Research Association and Science Fiction Writers of America.
11. Niedzviecki, *Trees On Mars,* 126.
12. Future of Humanity Institute, University of Oxford, https://www.fhi.ox.ac.uk/.

Chapter 1: Isaac's Kids (1950)

1. "Timeline," *Bulletin of the Atomic Scientists*, http://thebulletin.org/timeline.
2. Spencer R. Weart, *The Rise of Nuclear Fear* (Massachusetts and London, England: Harvard University Press, 2012), 106.
3. Weart, *The Rise of Nuclear Fear*. 73.
4. Ibid. Page 74. Quoting from *"Please don't let them": SANE Education Fund, "Shadows of the Nuclear Age: American Culture and the Bomb" (WGBH-FM broadcast and cassettes, 1980), cassette 4.*
5. John R. Platt, "Books That Make a Year's Reading and a Lifetime's Enrichment," *New York Times*, February 2, 1964.
6. Jeremy Norman, HistoryofInformation.com, October 9, 2016.
7. Markoff, *Machines of Loving Grace*, 98–99.
8. Markoff, *Machine of Loving Grace*, 104.
9. Associated Press, "Marvin Minsky: A pioneer of artificial intelligence." *The Globe and Mail*, (Toronto), January 27, 2016.
10. Ray Kurzweil, *The Singularity Is Near: When Humans Transcend Biology* New York: Viking, New York, 2005, 2.
11. Jeremy Pearce, "Joseph Engelberger: A Leader of the Robot Revolution," *The Globe and Mail*, (Toronto), December 8, 2015.
12. Michael White, *Isaac Asimov: A Life of the Grand Master of Science Fiction* (New York: Carrol & Graf, 2005), 52–55.
13. White, *Isaac Asimov: A Life of the Grand Master*, 53.
14. Bill Bryson, *At Home: A Short History of Private Life* (Anchor Canada, 2010), 157.
15. White, *Isaac Asimov: A Life of the Grand Master*, Pages 52–55.
16. Minsoo Kang, *Sublime Dreams of Living Machines: The Automaton in European Imagination* (Boston: Harvard University Press, 2011), 23–24.
17. Isaac Asimov, "Runaround," *Astounding*, March 1942.
18. White, *Isaac Asimov: A Life of the Grand Master*, 56.
19. *60 Minutes*, October 9, 2016.
20. "Our Friend The Atom," *Walt Disney's Disneyland*, Air date: January 23, 1957, https://www.youtube.com/watch?v=QDcjW1XSXN0.
21. Norbert Wiener, *The Human Use of Human Beings* (New York: Houghton Mifflin, 1950), Page
22. "Mars And Beyond," *Walt Disney's Disneyland*, Air date: December 4, 1957, https://www.youtube.com/watch?v=iEg7dF5rg8Y.

23. Elon Musk, AllThingsD conference, 2013.
24. *60 Minutes*, October 9, 2016.

Chapter 2: Monolith (1968)

1. Mark Aldrich, "History of Workplace Safety in the United States: 1880 to 1970," Website of the Economic History Association, eh.net.
2. Hawkins and Staff, *Hawkins Electrical Guide Number One: A Progressive course of Study for Engineers, Electricians, Students and Those Desiring To Acquire A Working Knowledge of Electricity and Its Applications* (New York: Theo. Audel & Co., 1917), Introductory Chapter.
3. Arthur C. Clarke, *2001: A Space Odyssey, Based on a Screenplay by Stanley Kubrick and Arthur C. Clarke*, 1999 edition, Originally published 1968 (New York: Roc, 1999), Introduction, viii.
4. Peter Kramer, *Film Classics: 2001: A Space Odyssey* (London: Palgrave MacMillan, 2010), 29.
5. Kramer, *Film Classics: 2001*, 19.
6. Clarke, *2001: A Space Odyssey*, xii.
7. Clarke, *2001: A Space Odyssey*, xiv.
8. Clarke, *2001: A Space Odyssey*, 92.
9. Clarke, *2001: A Space Odyssey*, 93–94.
10. Ana Matronic, *Robot Universe: Legendary Automatons and Androids from the Ancient World to the Distant Future* (New York: Sterling Publishing, 2015), 179.
11. Matronic, *Robot Universe*, 179.
12. Clarke, *2001: A Space Odyssey*, 125.
13. Walter Black (story) and Tony Benedict (teleplay), "Rosey's Boyfriend," TV episode, *The Jetsons*, Season 1. 1962.
14. Barry Blitzer (story) and Tony Benedict (teleplay), "Uniblab," TV episode, *The Jetsons*, Season 1. 1962.
15. Andreas Bernard, *Lifted: A Cultural History of the Elevator* (New York and London: New York University Press, 2014).
16. Bernard Darrach, "Meet Shaky [sic], The First Electronic Person," *Life* Magazine, November 20, 1970, Quoted in cyberneticzoo.com/cyberneticanimals/1967-shakey-charles-rosen-nils-nilsson-bertram-raphael-et-al-american/.
17. Markoff, *Machines of Loving Grace*, Page 195.
18. Markoff, Machines of Loving Grace.
19. https://en.wikipedia.org/wiki/Lights_out_(manufacturing).

Chapter 3: A Tramp in the AI Winter (1985)

1. Jane Caputi, http://xroads.virginia.edu/~drbr/caputi.html.
2. "Letters to PC," *PC: The Independent Guide to IBM Personal Computers,* Charter Issue, 1982, 14.
3. Roger Rosenblatt, "A New World Dawns," *TIME* Magazine, January 3, 1983.
4. Christina Bonnington, "The Modern PC Turns 30," *WIRED*, December 12, 2011.
5. Jeremy Joan Hewes, "PC for a Publisher," *PC: The Independent Guide to IBM Personal Computers,* Charter Issue, 1982, 68.
6. David Bunnell, "The Man Behind The Machine?" *PC: The Independent Guide to IBM Personal Computers,* Charter Issue, 1982, 21.
7. https://www.wired.com/2014/11/ferris-bueller-movie-gadgets/#slide-4.
8. Markoff, *Machines of Loving Grace,* 133–134.
9. *Machines of Loving Grace,* 140.
10. http://police-technology.net/id55.htm.
11. Michael Swaine and Paul Freiberger, *Fire in the Valley (Third Edition): The Birth and Death of the Personal Computer* (Dallas, Texas and Raleigh, North Carolina: Pragmatic Bookshelf, 2014), 11.
12. Swaine and Freiberger, *Fire in the Valley, 262–269.*
13. Swaine and Freiberger, *Fire in the Valley.*
14. Walter Isaacson, *Steve Jobs.* (New York: Simon & Shuster, 2011). 97.
15. Isaacson, *Steve Jobs.*
16. James Carter, http://www.starringthecomputer.com/.
17. Swaine and Freiberger, *Fire in the Valley, 38–41.*
18. Swaine and Freiberger, *Fire in the Valley,*
19. Swaine and Freiberger, *Fire in the Valley, 30.*
20. Steve Wozniak, Atariarchives.org.
21. Wozniak, Atariarchives.org.
22. Swaine and Freiberger, *Fire in the Valley,* 298.
23. Swaine and Freiberger, *Fire in the Valley,* 297.
24. Swaine and Freiberger, *Fire in the Valley,* 355.
25. http://www.history.com/topics/inventions/automated-teller-machines.
26. http://www.history.com/topics/inventions/automated-teller-machines.

Chapter 4: Fun and Games with Thinking Machines (2018)

1. Markoff, *Machines of Loving Grace,* 109.
2. Garry Kasparov, "The Day That I Sensed A New Kind of Intelligence," *TIME,* March 25, 1996.

3. Jennifer Latson, "Did Deep Blue Beat Kasparov Because Of A System Glitch?" *TIME*, Feb 17, 2015. http://time.com/3705316/deep-blue-kasparov/.
4. Latson, "Did Deep Blue Beat Kasparov?"
5. Brian Jenkins, "Chess fans overload IBM's website," CNN.com, May 7, 1977, http://www.cnn.com/WORLD/9705/03/chess.rematch/index.html.
6. IBM100, "Icons of Progress: A Computer Called Watson," http://www-03.ibm.com/ibm/history/ibm100/us/en/icons/watson/.
7. Vikram Jayanti, Director, *Game Over; Kasparov And The Machine*, (Documentary film), Canada/United Kingdom, 2003.
8. Jayanti, *Game Over*.
9. Jayanti, *Game Over*.
10. Jayanti, *Game Over*.
11. Julia Cort & Michael Bicks, writers; Michael Bicks, director, PBS, *NOVA* documentary: "IBM Watson: Smartest Machine On Earth," February 2011.
12. Ken Jennings, "My Puny Human Brain," February 16, 2011, Slate.com.
13. Cort and Bicks, "Smart Machine on Earth."
14. Liat Clark, "DeepMind's AI is an Atari gaming pro now," WIRED, February 25, 2015, http://www.wired.co.uk/article/google-deepmind-atari.
15. Cort and Bicks, "Smart Machine on Earth."
16. Michael Neilson, "Using neural nets to recognize handwritten digits," *Neural Networks and Deep Learning*, (January 2017), http://neuralnetworksanddeeplearning.com/chap1.html.
17. Jo Best, "IBM Watson: The inside story of how the Jeopardy-winning supercomputer was born and what it wants to do next," *Tech Republic*, 2016.
18. Cort and Bicks, "Smart Machine on Earth."
19. Jennings, "My Puny Human Brain."
20. Jennings, "My Puny Human Brain."
21. Jo Best, "IBM Watson."
22. Amara D. Angelica, Editor, *Kurzweil Accelerating Intelligence*, Blog, http://www.kurzweilai.net/the-buzzer-factor-did-watson-have-an-unfair-advantage, Feb. 24, 2011.
23. Cade Metz, "Google's Go Victory Is Just A Glimpse of How Powerful AI Will Be," *WIRED*, January 2016.
24. Shelly Palmer, "AlphaGo vs. You: Not a Fair Fight," March 13, 2016, www.shellypalmer.com.

25. Palmer, "AlphaGo vs. You."
26. Palmer, "AlphaGo vs. You."
27. Byron Spice, Carnegie Mellon University Communications Department press release, "Computer Out-Plays Humans in "Doom," September 27, 2016.
28. Chris Wilson, "Jeopardy, Schmeopardy: Why IBM's next target should be a machine that plays poker," Feb. 15, 2011, Slate.com.
29. https://www.riverscasino.com/pittsburgh/BrainsVsAI/.
30. Liat Clark, "Google's Artificial Brain Learns to Find Cat Videos," *WIRED*, June, 2016, https://www.wired.com/2012/06/google-x-neural-network/.
31. David C. Robertson with Bill Breen, *Brick by Brick: How LEGO Rewrote the Rules of Innovation and Conquered the Global Toy Industry* (New York: Crown Publishing Group, 2013), 181–185.
32. Cort and Bicks, "Smart Machine on Earth."
33. Garry Kasparov, "The Chess Master and the Computer," *The New York Review of Books*, February 11, 2010.

Chapter 5: Hitching A Ride In A Driverless Car (2025)

1. National Safety Council, http://www.nsc.org/learn/pages/nsc-on-the-road.aspx?var=hpontheroad.
2. Reese, "A List Of The World's Self-Driving Cars Racing Toward 2020."
3. John Jordan, "Robots," *The MIT Press*, (Cambridge, Massachusetts: 2016), 100.
4. J. D. Power, http://www.jdpower.com/press-releases/jd-power-2017-us-tech-choice-study.
5. "Uber's problems could spark ride-hailing customer grab," *TU Automotive Week Brief*, June 26, 2017.
6. Greg Keenan, "Electric vehicles not expected to rule the road," *The Globe & Mail*, Toronto, June 14, 2017.
7. Serge Lazarro, "Self-Driving Cars Will Cause Motion Sickness 'Often' to 'Always,' Study Finds," *Observer*, June 2, 2015, http://observer.com/2015/06/self-driving-cars-will-cause-motion-sickness-often-to-always-study-finds/.
8. Hod Lipson and Melba Kurman, *Driverless: Intelligent Cars and the Road Ahead*. (Cambridge, Massachusetts: The MIT Press, 2016), 48.
9. Jordan, "Robots," 106.
10. Lipson and Kurman, *Driverless*, 79.
11. Markoff, *Machines of Loving Grace*, 113–114.

12. Markoff, *Machines of Loving Grace*, 21.
13. Jordan, "Robots," 105.
14. Jordan, "Robots," 102–106.
15. Andrew J. Hawkins, "Apple just received a permit to test self-driving cars in California," *The Verge*, April 14, 2017, https://www.theverge.com/2017/4/14/15303338/apple-autonomous-vehicle-testing-permit-california.
16. Kristin Lee, "How the cars of Logan grappled with the very real future," *Jalopnik*, March 10, 2017, http://jalopnik.com/how-the-cars-of-logan-grappled-with-the-very-real-futur-1793099275.
17. *TU Auto Weekly Brief*, "Connectivity is road to revenues says Harman," June 2, 2017.
18. *TU Automotive Detroit 2017 Weekly Brief*, June 12, 2017.
19. Marcus Berret et al., *Automotive Disruption Radar*: Issue #1, Roland Berger GMBH, (Munich, 2017), Pages 10–11.
20. Matt Bubbers, Jordan Chittley, and Mark Richardson, "The Future of Mobility," *The Globe and Mail*, January 5, 2017.
21. Lipson and Kurman, *Driverless*: Intelligent Cars and the Road Ahead, 98–106. The MIT Press. (Cambridge, Massachusetts: 2016). Pages 98 to 106.
22. Ashlee Vance, "The First Person to Hack the iPhone Built A Self-Driving Car In His Garage," *Bloomberg Business*, December 16,2015; Tesla Motors, correction to article, Quoted by Hod Lipson and Melba Kurman, *Driverless: Intelligent Cars and the Road Ahead*, (Cambridge, Massachusetts: The MIT Press, 2016), 86.
23. Janet W. Burns, "Japanese Leaders Aim to Make Tokyo A Self-Driving City for 2020 Olympics," *Forbes.com*, September 8, 2016, https://www.forbes.com/sites/janetwburns/2016/09/08/japanese-leaders-aim-to-make-tokyo-a-self-driving-city-for-2020-olympics/#61cf96f81090.

Chapter 6: Waking Up the House (2030)

1. Jill Lepore, "The Cobweb," *The New Yorker*, January 26, 2015, http://www.newyorker.com/magazine/2015/01/26/cobweb.
2. Lepore, "The Cobweb."
3. Thien-Kim Lam, "14 Sex Toys You Can Control with your Smartphones," Momtastic blog, www.ca.momtastic.com/love-sex.
4. http://www.postscapes.com/connected-kitchen-products/#flatware.
5. *Waking Up with Sam Harris*, Podcast #66, Interview with researcher Kate Darling of MIT Media Lab, March 1, 2017.

6. Kevin Ashton, http://www.rfidjournal.com/articles/view?4986.
7. http://fusion.net/robots-dream-up-hilarious-cooking-recipes-1793852524.
8. https://www.ibmchefwatson.com/community.
9. Niedzviecki, *Trees On Mars,* 204.
10. Markoff, *Machines of Loving Grace,* 114.
11. Eustacia Huen, "The World's First Home Robotic Chef Can Cook Over 100 Meals," www.forbes.com, October 31, 2016.
12. http://www.webmd.com/beauty/cosmetic-procedures-overview-skin#1.
13. http://www.livescience.com/33179-does-human-body-replace-cells-seven-years.html.
14. Kevin Kelly, *The Inevitable: Understanding the 12 Technological Forces That Will Shape Our Future* (New York: Viking, 2016), 10–11.
15. Markoff, *Machines of Loving Grace,* 201–203.
16. Rodney Brooks, "Robots will invade our lives," *TED Talk,* https://www.ted.com/talks/rodney_brooks_on_robots February 2003.
17. http://www.irobot.com/About-iRobot/STEM/Create-2.aspx.
18. http://media.irobot.com/2017-03-15-iRobot-Takes-Next-Step-in-the-Connected-Home-with-Clean-Map-TM-Reports-and-Amazon-Alexa-Integration.
19. http://www.theglobeandmail.com/life/home-and-garden/gardening/robomowers-let-you-sit-back-and-watch-the-grass-go/article31137694/.
20. http://www.rethinkrobotics.com/baxter/.
21. https://www.nytimes.com/2016/06/23/technology/personaltech/mark-zuckerberg-covers-his-laptop-camera-you-should-consider-it-too.html?_r=0.
22. Andrea Peterson, "Can anyone keep us safe from a weaponized 'Internet of Things'?" *The Washington Post,* October 25, 2016.
23. Peter Novak, "How safe is the Internet of Things?" December 6, 2015, thestar.com.
24. Novak, "How safe is the Internet of Things?"
25. Dan Meth, "Welcome to Uncanny Valley . . . where robots make you want to throw up," *BuzzFeed,* August 19, 2014.
26. *Waking Up with Sam Harris,* Podcast #66.

Chapter 7: The Good Robot (2040)

1. Luke Edwards, Pocket Lint, "What is Elon Musk's 700 mph Hyperloop? The subsonic train explained," May 10, 2016, http://www.pocket-lint.com/news/132405-what-is-elon-musk-s-700mph-hyperloop-the-subsonic-train-explained.

2. Malcolm James, "Here's how Bill Gates' plan to tax robots could actually happen," *Business Insider,* March 20, 2017, http://www.businessinsider.com/bill-gates-robot-tax-brighter-future-2017-3.
3. Evan Ackerman, "Researchers teaching robots to feel and react to pain," *IEEE Spectrum,* May 24, 2016, http://spectrum.ieee.org/automaton/robotics/robotics-software/researchers-teaching-robots-to-feel-and-react-to-pain.
4. Leslie Katz, "Walk With Me: Robotic exosuits are giving paraplegics a new view of the world around them," *CNET,* Spring 2017 issue.
5. Lauren Orsini, "Jibo's Cynthia Breazeal: Why we will learn to love our robots," *Readwrite,* August 11, 2014. http://readwrite.com/2014/08/11/jibo-cynthia-breazeal-robots-social/.
6. Chris Atkeson, "What the future of robots could look like," CNN.com. December 27, 2014. http://www.cnn.com/2014/12/27/opinion/atkeson-soft-robots-care/index.html.
7. Brian Clegg, *Ten Billion Tomorrows: How Science Fiction Technology Became Reality And Shapes The Future* (London : St. Martin's Press, 2015), 67.
8. Julian Robinson. "The world's first robot 'actress': Talking android fitted with a human face is given star role in Japanese nuclear disaster film," *Daily Mail Online* (UK), November 1, 2015, http://www.dailymail.co.uk/sciencetech/article-3299487/The-world-s-robot-actress-Talking-android-fitted-human-face-given-star-role-Japanese-nuclear-disaster-film.html.
9. Jordan, "Robots," 202–204.
10. https://en.wikipedia.org/wiki/Astro_Boy.
11. *Astro Boy* TV show, Episode 1: "The Birth of Astro Boy," Nippon Television Corp., Tezuka Production Inc., Created by Osama Tezuka, 1982; https://www.youtube.com/watch?v=Vjck3XuNqn4.
12. Felix Salmon and John Stokes, "Algorithms Take Control of Wall Street," *WIRED,* December 27, 2010. https://www.wired.com/2010/12/ff_ai_flashtrading/.
13. Rob Price, "Microsoft is deleting its AI chatbot's incredibly racist tweets," *Business Insider UK,* March 24, 2016, http://uk.businessinsider.com/microsoft-deletes-racist-genocidal-tweets-from-ai-chatbot-tay-2016-3.

Chapter 8: Sex and the Singularity (2050)

1. ZeeNewsIndia, published by Sandra Hensel, "3D-Printed Skin Could Allow Robots To 'Feel,'" Robouniverse.com, May 15, 2017, http://robouniverse.com/news/3d-printed-skin-could-allow-robots-to-feel/34959/.

2. Kurzweil, *The Singularity Is Near*. 301.
3. Jaron Lanier, *You Are Not A Gadget: A Manifesto* (New York: Alfred A. Knopf, 2010).
4. Kurzweil, *The Singularity Is Near*, 205.
5. Kurzweil, *The Singularity Is Near*, 10–11.
6. Kurzweil, *The Singularity Is Near*.
7. David Levy, *Love and Sex With Robots: The Evolution of Human-Robot Relationships*, (New York: Harper Collins, 2007), 177.
8. Levy, *Love and Sex With Robots*, 236–237.
9. Levy, *Love and Sex With Robots*, 179.
10. Levy, *Love and Sex With Robots*, 257–259.
11. Levy, *Love and Sex With Robots*, 253–256.
12. Levy, *Love and Sex With Robots*, 262–265.
13. Levy, *Love and Sex With Robots*, 248–249.
14. Cliff Lee, "Who still hangs out on Second Life? More than half a million people," *The Globe & Mail*, May 17, 2017, https://www.theglobeandmail.com/life/relationships/who-still-hangs-out-on-second-life-more-than-half-a-million-people/article35019213/.
15. Ramna Shahzad, "Online dating is exhausting so this woman got a robot to swipe and choose men for her," *CBC News*, www.cbc.ca, May 24, 2017. http://www.cbc.ca/news/canada/toronto/online-dating-1.4129820
16. Levy, *Love and Sex With Robots*, 38–41.
17. Levy, *Love and Sex With Robots*, 33.
18. Levy, *Love and Sex With Robots*, 243.
19. www.realdoll.com.
20. Emma Cott, "Sex Dolls That Talk Back," *The New York Times*, June 11, 2015.
21. Christopher Trout, "RealDoll's First Sex Robot Took Me to the Uncanny Valley," www.engadget.com, April 11, 2017, https://www.engadget.com/2017/04/11/realdolls-first-sex-robot-took-me-to-the-uncanny-valley/.
22. David Levy interviewed by Brent Bambury on "Day 6," CBC Radio. January 6, 2017, http://www.cbc.ca/radio/day6/episode-319-becoming-kevin-o-leary-saving-shaker-music-google-renewables-marrying-robots-and-more-1.3921088/a-i-expert-david-levy-says-a-human-will-marry-a-robot-by-2050-1.3921101.
23. ZeeNewsIndia, "3D-Printed Skin."
24. Trout, "RealDoll's."
25. Matronic, Robot Universe, 157.

26. Tim Minnear, Scriptwriter, Joss Whedon's *Firefly*, "Out Of Gas," Episode aired November 25, 2002.
27. Vernor Vinge, "The Coming Singularity: How To Survive in the Post-Human Era," Article for the VISION-21 Symposium sponsored by NASA Lewis Research Center and the Ohio Aerospace Institute, March 30-31, 1993, https://archive.org/stream/pdfy-MZnoomxoY3Kv2X9O/Vernor Vinge The coming technological singularity_djvu.txt.
28. Vinge, "The Coming Singularity."
29. Kurzweil, *The Singularity Is Near*, 309.
30. Ben Popper, "Rapture of the nerds: Will the Singularity turn us into gods or end the human race?" *The Verge*, October 22, 2012, https://www.theverge.com/2012/10/22/3535518/singularity-rapture-of-the-nerds-gods-end-human-race.
31. Darrell Etherington, "Elon Musk could soon share more on his plan to help humans keep up with AI," *Tech Crunch*, January 25, 2017. https://techcrunch.com/2017/01/25/elon-musk-could-soon-share-more-on-his-plan-to-help-humans-keep-up-with-ai/.
32. The Economist, "The case for neural lace: Elon Musk enters the world of brain-computer interfaces," March 30, 2017. http://www.economist.com/news/science-and-technology/21719774-do-human-beings-need-embrace-brain-implants-stay-relevant-elon-musk-entershttps://www.recode.net/2016/6/2/11837544/elon-musk-neural-lace.
33. Greg Satell, "3 Reasons to Believe the Singularity is Near," Forbes.com. June 3, 2016.
34. Markoff, *Machines of Loving Grace*, 84–94.
35. Simon Pegg and Edgar Wright, *The World's End*, 2013.
36. Kurzweil, *The Singularity Is Near*, 24.
37. Jason Dorrier, "Looking Ahead As Moore's Law Turns 50: What's Next For Computing?" *Singularity Hub*, April 20, 2015, https://singularityhub.com/2015/04/20/looking-ahead-as-moores-law-turns-50-whats-next-for-computing/.
38. Alex Garland, Scriptwriter, *Ex Machina*, 2014.
39. Garland, *Ex Machina*.
40. Isaac Asimov interview with Bill Moyers, 1988, https://www.youtube.com/watch?v=Uj7IXIAReMM.
41. Isaac Asimov, Robert Silverberg, and Nicholas Kazan, *Bicentennial Man*, (Film), 1999.
42. Isaac Asimov. *Foundation and Earth*, (New York: Doubleday, 1986).